Multi-Walled Carbon Nanotubes

Multi-Walled Carbon Nanotubes

Special Issue Editor

Simone Morais

MDPI • Basel • Beijing • Wuhan • Barcelona • Belgrade

MDPI

Special Issue Editor
Simone Morais
Instituto Superior de Engenharia do Porto
Portugal

Editorial Office
MDPI
St. Alban-Anlage 66
4052 Basel, Switzerland

This is a reprint of articles from the Special Issue published online in the open access journal *Applied Sciences* (ISSN 2076-3417) from 2018 to 2019 (available at: https://www.mdpi.com/journal/applsci/special_issues/Multi-Walled_Carbon_Nanotubes)

For citation purposes, cite each article independently as indicated on the article page online and as indicated below:

LastName, A.A.; LastName, B.B.; LastName, C.C. Article Title. *Journal Name* **Year**, *Article Number*, Page Range.

ISBN 978-3-03921-229-3 (Pbk)
ISBN 978-3-03921-230-9 (PDF)

Contents

About the Special Issue Editor

Simone Morais has a Ph.D. (1998) in chemical engineering from the University of Porto. She is an associate professor in the Department of Chemical Engineering at the School of Engineering of the Polytechnic of Porto (Portugal) and a permanent researcher at REQUIMTE–LAQV (http://www.requimte.pt/laqv/). REQUIMTE–LAQV is a science-driven institution focused on the development of sustainable chemistry. Her current research interests are chemically modified electrodes, electroanalysis, (bio)sensors, the preparation and application of nanofunctional materials, and new methodologies for environmental pollutant analysis. She has co-authored about 140 papers (ORCID: 0000-0001-6433-5801; Scopus ID 7007053747) in journals with impact factors and about 30 book chapters. She has also supervised several Ph.D. and post-doctoral fellows and has participated in and coordinated several projects.

*applied
sciences*

MDPI

Editorial

Multi-Walled Carbon Nanotubes

Simone Morais

REQUIMTE–LAQV, Instituto Superior de Engenharia do Porto, Instituto Politécnico do Porto, Rua Dr.
Bernardino de Almeida 431, 4249-015 Porto, Portugal; sbm@isep.ipp.pt; Tel.: +351-228340500;
Fax: +351-228321159

Received: 30 June 2019; Accepted: 1 July 2019; Published: 2 July 2019

1. Introduction

Since their discovery, multi-walled carbon nanotubes (MWCNTs) have received tremendous attention because of their unique electrical, optical, physical, chemical, and mechanical properties [1]. Their particular characteristics make them well-matched for a plethora of application areas, namely, nanoelectronics, energy management, (electro)catalysis, materials science, the construction of (bio)sensors, multifunctional nanoprobes for biomedical imaging, sorbents for sample preparation, the removal of contaminants from wastewater, as anti-bacterial agents, drug delivery nanocarriers, and so on—the current relevant application areas are countless. This Special Issue is a collection of 13 original research articles that address remarkable advances in the synthesis, purification, characterization, functionalization, and application of MWCNTs in established and emerging areas. A brief discussion of the main outcomes of each study is presented in the next sections.

2. Synthesis and Structural Characterization of Multi-Walled Carbon Nanotubes-Based (Nano)Composites

The trends and advances regarding the synthetic routes and structural properties of MWCNTs-based (nano)composites have been discussed in several reports [2–6], proving the importance of this topic for pushing MWCNT exploitation and industrial use forward. Ma et al. [2] proposed a new route for the synthesis of MWCNT nanohybrids using azide-terminated poly(methyl methacrylate), through the utilization of a combination of reversible addition fragmentation chain transfer and the alkyne-azide click reaction. The as-prepared nanohybrids could be steadily dispersed in aqueous solutions, including water, because of the azide-terminated poly(methyl methacrylate) chain on the MWCNT surface. Chandra et al. [3] inserted nanopalladium on carboxylated and octadecylamine functionalized MWCNTs that were further extensively characterized. The synthesized hybrids exhibited a good catalytic activity towards a carbon–carbon coupling reaction. Liu et al. [4] and Savi et al. [5] synthesized and characterized new composites using silicone rubber/polyolefin elastomer blends containing ionic liquids modified with carbon blacks and MWCNTs, and MWCNTs dispersed in an epoxy resin matrix, respectively. The prepared composites have an interesting potential for electromagnetic interference shielding applications [4] and microwave absorbing uses (although only low weight percentage should be used) [5]. In addition, Min et al. [6] developed a novel nanocomposite based on acidified MWCNTs, graphene oxide, and cerium oxide nanoparticles, which displayed a promising tribological performance.

3. Modelling of Multi-Walled Carbon Nanotubes-Based Nanofluid Flow

Nanofluids are obtained by the incorporation of nanomaterials into a base fluid, with the main goal of enhancing specific properties, such as the density, viscosity, thermal conductivity, and specific heat. Thus, they have been increasingly developed and characterized, using several mathematical models, for further applications, mainly in the field of engineering (heat interchangers, freezing, cooling, heating systems, etc.). Muhammad et al. [7] described the three-dimensional rotational flow of three different MWCNT-based nanofluids (prepared with water, engine oil, and kerosene oil as the base liquid) considering thermal radiation and heat generation/absorption. The main parameters of the non-Newtonian behavior of the several nanofluids were determined and discussed. Saba et al. [8] also studied the flow of nanofluids composed of water and MWCNTs. The authors of [8] thoroughly investigated the several variables that affect the MWCNT-based nanofluid flow over a curved stretching surface and for heat transfer distribution.

4. Applications

4.1. Adsorption

MWCNTs are also being successfully explored in environmental applications, such as for water quality control and treatment. Two interesting reports [9,10] are included in this Special Issue, both of which are related to the use of MWCNTs as adsorbents of contaminants, namely nonylphenol [9] and organochlorine pesticides [10] from source waters and agricultural irrigation water samples, respectively. The contributing authors [9,10] characterized the main operational parameters and the type of adsorption demonstrating the applicability of MWCNTs for the removal and extraction of contaminants from water samples.

Alguacil [11] also characterized the possibility of applying MWCNTs as sorbents, but in this case, for gold(I) and gold(III) cations' adsorption from cyanide and chloride solutions. The reached data suggested that the recovery of the selected metal may be accomplished by subsequent elution with acidic thiourea solutions (for cyanide medium) or with aqua regia (for chloride solutions), with a further possibility of obtaining zero-valent gold nanoparticles.

4.2. Sensors Design

MWCNTs have been extensively incorporated into electrochemical (bio)sensors' design, regardless of the detection scheme and the target analyte, because of their inherent properties. Their high conductivity, catalytic properties, high surface area, chemical stability, and biocompatibility promote a significant increase in the sensitivity, lifetime, and overall performance of the devices, as concluded in the contribution of Oliveira et al. [12]. Nevertheless, the authors of [12] also concluded, after analyzing the published data from the 2013–2018 period concerning the new generation of sensors, that further technical developments are still needed in order to lower the cost of production of high quality MWCNTs. Moreover, the lack of comprehensive characterization of the toxicity of MWCNTs was also identified as an issue for the increase of the in vivo MWCNTs based sensors usage.

4.3. Drug Delivery

MWCNTs (functionalized or not) have been exploited in the drug delivery, biochemistry, and medicine fields. Chen et al. [13] described the effect of different levels of the carboxylation of MWCNTs on the dissolution rate of sulfamethoxazole and griseofulvin, two therapeutic hydrophobic drugs used as antibiotic and antifungal agents, respectively. The anti-solvent synthesis of micron-scale drug particles was the applied technique [13]. The reached data suggested that the degree of functionalization may help to control the release of the drugs, as decreasing the C:COOH ratio in the functionalized MWCNTs promoted a significant increase in the dissolution rates [13].

4.4. Cementitious Materials

The incorporation of nanomaterials including MWCNTs in building materials is increasingly being characterized in order to enhance the mechanical, physical, and electrical properties of the structures, while reducing their failure. Dalla et al. [14] studied the influence of introducing MWCNTs as nano-reinforcements in cement mortars. The obtained results showed that the permeability, electrical resistivity, and the flexural and compressive properties of the mortars were significantly affected by the inclusion of MWCNTs at levels ranging from 0.2–0.8 wt % of cement [14].

Funding: I am grateful for the financial support from the European Union (FEDER funds through COMPETE) and National Funds (Fundação para a Ciência e Tecnologia—FCT), through projects UID/QUI/50006/2019 and PTDC/ASP-PES/29547/2017 (POCI-01-0145-FEDER-029547) by FCT/MEC, with national funds and co-funded by FEDER.

Acknowledgments: All contributing authors and reviewers, as well as the technical support of the editorial team of Applied Sciences (in particular Emily Zhang) are greatly acknowledged. I sincerely thank all of them for their hard work and for the opportunity to work with them in this Special Issue. I also wish that readers from the different research fields will enjoy and find useful this Open Access Special Issue.

Conflicts of Interest: The author declares no conflict of interest.

References

1. Soriano, M.S.; Zougagh, M.; Valcárcel, M.; Ríos, Á. Analytical nanoscience and nanotechnology: Where we are and where we are heading. *Talanta* **2018**, *177*, 104–121. [CrossRef] [PubMed]

2. Ma, W.; Zhao, Y.; Zhu, Z.; Guo, L.; Cao, Z.; Xia, Y.; Yang, H.; Gong, F.; Zhong, J. Synthesis of poly(methyl methacrylate) grafted multiwalled carbon nanotubes via a combination of RAFT and alkyne-azide click reaction. *Appl. Sci.* **2019**, *9*, 603. [CrossRef]

3. Chandra, B.; Wu, Z.; Ntim, S.; Rao, G.; Mitra, S. The effect of functional group polarity in palladium immobilized multiwalled carbon nanotube catalysis: Application in carbon–carbon coupling reaction. *Appl. Sci.* **2018**, *8*, 1511. [CrossRef] [PubMed]

4. Liu, C.; Yu, C.; Sang, G.; Xu, P.; Ding, Y. Improvement in EMI shielding properties of silicone Rubber/POE blends containing ILs modified with carbon black and MWCNTs. *Appl. Sci.* **2019**, *9*, 1774. [CrossRef]

5. Savi, P.; Giorcelli, M.; Quaranta, S. Multi-walled carbon nanotubes composites for microwave absorbing applications. *Appl. Sci.* **2019**, *9*, 851. [CrossRef]

6. Min, C.; He, Z.; Song, H.; Liu, D.; Jia, W.; Qian, J.; Jin, Y.; Guo, L. Fabrication of novel CeO_2/GO/CNTs ternary nanocomposites with enhanced tribological performance. *Appl. Sci.* **2019**, *9*, 170. [CrossRef]

7. Muhammad, S.; Ali, G.; Shah, Z.; Islam, S.; Hussain, S. The rotating flow of magneto hydrodynamic carbon nanotubes over a stretching sheet with the impact of non-linear thermal radiation and heat generation/absorption. *Appl. Sci.* **2018**, *8*, 482. [CrossRef]

8. Saba, F.; Ahmed, N.; Hussain, S.; Khan, U.; Mohyud-Din, S.; Darus, M. Thermal analysis of nanofluid flow over a curved stretching surface suspended by carbon nanotubes with internal heat generation. *Appl. Sci.* **2018**, *8*, 395. [CrossRef]

9. Dai, Y.; Shah, K.; Huang, C.; Kim, H.; Chiang, P. Adsorption of nonylphenol to multi-walled carbon nanotubes: Kinetics and isotherm study. *Appl. Sci.* **2018**, *8*, 2295. [CrossRef]

10. Huang, X.; Liu, G.; Xu, D.; Xu, X.; Li, L.; Zheng, S.; Lin, H.; Gao, H. Novel zeolitic imidazolate frameworks based on magnetic multiwalled carbon nanotubes for magnetic solid-phase extraction of organochlorine pesticides from agricultural irrigation water samples. *Appl. Sci.* **2018**, *8*, 959. [CrossRef]

11. Alguacil, F. Adsorption of gold(I) and gold(III) using multiwalled carbon nanotubes. *Appl. Sci.* **2018**, *8*, 2264. [CrossRef]

12. Oliveira, T.; Morais, S. New generation of electrochemical sensors based on multi-walled carbon nanotubes. *Appl. Sci.* **2018**, *8*, 1925. [CrossRef]

13. Chen, K.; Mitra, S. Controlling the dissolution rate of hydrophobic drugs by incorporating carbon nanotubes with different levels of carboxylation. *Appl. Sci.* **2019**, *9*, 1475. [CrossRef]

14. Dalla, P.; Tragazikis, I.; Exarchos, D.; Dassios, K.; Barkoula, N.; Matikas, T. Effect of carbon nanotubes on chloride penetration in cement mortars. *Appl. Sci.* **2019**, *9*, 1032. [CrossRef]

Article

Synthesis of Poly(methyl methacrylate) Grafted Multiwalled Carbon Nanotubes via a Combination of RAFT and Alkyne-Azide Click Reaction

Wenzhong Ma [1,*], Yuchen Zhao [1], Zhiwei Zhu [1], Lingxiang Guo [1], Zheng Cao [1], Yanping Xia [1], Haicun Yang [1], Fanghong Gong [1] and Jing Zhong [2]

[1] Jiangsu Key Laboratory of Environmentally Friendly Polymeric Materials, School of Materials Science and Engineering, Jiangsu Collaborative Innovation Center of Photovoltaic Science and Engineering, Changzhou University, Changzhou 213164, Jiangsu, China; zhaoyuchentmac@foxmail.com (Y.Z.); zedzzw@gmail.com (Z.Z.); glx16441216@gmail.com (L.G.); zcao@cczu.edu.cn (Z.C.); xiayanping0715@126.com (Y.X.); yhcbobo@cczu.edu.cn (H.Y.); fhgong@cczu.edu.cn (F.G.)

[2] Jiangsu Key Laboratory of Advanced Catalytic Materials and Technology, School of Petrochemical Engineering, Changzhou University, Changzhou 213164, Jiangsu, China; zjwyz@cczu.edu.cn

* Correspondence: wenzhong-ma@cczu.edu.cn; Tel.: +86-519-86330095

Received: 28 December 2018; Accepted: 4 February 2019; Published: 12 February 2019

Abstract: An efficient synthesis route was developed for the preparation of multiwalled carbon nanotube (MWCNT) nanohybrids using azide-terminated poly(methyl methacrylate) (PMMA) via a combination of reversible addition fragmentation chain transfer (RAFT) and the click reaction. A novel azido-functionalized chain transfer agent (DMP-N₃) was prepared and subsequently employed to mediate the RAFT polymerizations of methyl methacrylate (MMA). The RAFT polymerizations exhibited first-order kinetics and a linear molecular weight dependence with the conversion. The kinetic results show that the grafting percentage of PMMA on the MWCNTs surface grows along with the increase of the reaction time. Even at 50 °C, the grafting rate of azide-terminated PMMA is comparatively fast in the course of the click reaction, with the alkyne groups adhered to MWCNTs in less than 24 h. The successful functionalization of PMMA onto MWCNT was proved by FTIR, while TGA was employed to calculate the grafting degree of PMMA chains (the highest GP = 21.9%). Compared with the pristine MWCNTs, a thicker diameter of the MWCNTs-*g*-PMMA was observed by TEM, which confirmed the grafted PMMA chain to the surface of nanotubes. Therefore, the MWCNTs-*g*-PMMA could be dispersed and stably suspended in water.

Keywords: multi-wall carbon nanotube (MWCNT); azide-alkyne click chemistry; RAFT polymerization; PMMA

1. Introduction

Nanoscience and nanotechnology have brought us the excellent development of many novel categories of functional materials and have become remarkable fields of study [1,2]. Recently, carbon nanotubes (CNTs) have acquired increasing importance and popularity in membrane science and technology due to their high permeability and selectivity, which they owe to the rapid flux through the hollow interior and nano-scale diameter of CNTs [3–7]. For instance, multiwalled carbon nanotubes (MWCNTs) with outer diameters (2–100 nm) exhibit a significantly high permeability in membrane process applications because of the large surface area [6]. The MWCNT hybrid nanostructure and composite materials with the introduction of polymer chemistry have dramatically attracted attention [8,9]. These nanocomposite materials complement the characteristics of functional polymers and thus provide improved nano-scale dispersing, hydrophilicity, electric properties, etc. [10]. Some researchers have directly immobilized MWCNTs into a polymeric membrane by the blending method due to its easy manipulation and mild conditions [11,12]. However, despite the outstanding properties of MWCNT composite materials, the tendency to be polymerized caused by the big inherent van der Waals forces of MWCNTs restricts its application to the fabrication of nanocomposites. [13].

Among the surface modification method, the "grafting to" technique is one of the most convenient techniques to cap polymer chains which can adjust the dispersibility of nanoparticles in polymer matrices [14]. In this approach, functional group-terminated polymer chains can graft onto the surface of nanoparticles in a highly efficient reaction, resulting in the formation of tethered polymer chains [15]. To achieve dense polymer layers attached to the MWCNT surface, strong interactions between polymer chains and the MWCNT surface are required. Poly(methyl methacrylate) (PMMA) is usually studied as a compatibilizer agent for polymer/nanoparticle composites [16]. Recently, "click chemistry" has attracted more attention in surface modification for nanoparticles due to its high yields without byproducts [17–20]. When combined with reversible addition-fragmentation chain transfer (RAFT), the precise predesign of the molecular weight, structure, and functionality of polymers can be controlled by living polymerizations [21]. For example, Singha and co-workers synthesized a hydrophilic MWCNT based upon the Diels-Alder (DA) click reaction by one step [22]. Nonetheless, few have been reported in the field of PMMA-functioned MWCNTs, synthesized through the click reaction.

In this work, by using azide/alkyne end groups, PMMA with an azide end group can effectively graft onto the MWCNT surface. To do this, the RAFT agent 2-dodecylsulfanylthiocarbonylsulfanyl-2-methylpropionic acid 3-azidopropyl ester (DMP-N$_3$) was used in the polymerization of PMMA. Then, azide-terminated PMMA was attached to the surface of alkyne-terminated MWCNTs via the "grafting to" approach. Due to the PMMA chain on the MWCNT surface, the nanohybrids can be stably dispersed in water and have potential for the preparation of MWCNT composite materials.

2. Materials and Methods

2.1. Materials

MWCNTs were obtained from Nanjing XFNANO Materials Tech Co., Ltd. (China). Methyl methacrylate (MMA) was obtained from Shanghai Chemical Plant (China). The inhibitor was removed by a basic alumina column, purified under lower pressure, and stored in an Ar atmosphere at −5 °C. Sodium azide, 4-(N,N-dimethylamino)pyridine(DMAP), and 1-(3-Dimethylaminopropyl)-3-ethylcarbodiimide were obtained from Aladdin Industrial Corporation (China). Azobisisobutyronitrile (AIBN) was purchased from Jiangsu Qiangsheng Chemical Co., Ltd. (China). N,N-dimethylacetamide (DMAc), propargyl alcohol, tetrahydrofuran (THF, analytical grade), and anisole (AR) were purchased from Shanghai Lingfeng Chemical Reagent Co., Ltd. (China). Thionyl chloride (analytical grade) was purchased from Sinopharm Chemical Reagent Co., Ltd. (China).

2.2. Synthesis of Azide-Terminated Poly(methyl methacrylate) (PMMA)

Before synthesizing azide-terminated PMMA, RAFT agent 2-dodecylsulfanylthiocarbonylsulfanyl-2-methylpropionic acid 3-azidopropyl ester (DMP-N$_3$) was synthesized according to the reported method [23]. FTIR analysis was performed to confirm successful azide-RAFT agent preparation, as shown in Figure 1a. FTIR (KBr) (wavenumber, cm^{-1}): 2923 (C-Cs), 2100 (C-N=N=N), 1735 (C=O), 1064 (C=S), 1250 (C-S). Figure 1b shows the ^{1}H NMR spectra of DMP-N$_3$. The peak at ^{1}H-NMR (400 MHz, CDCl$_3$, TMS) for DMP-N$_3$ (δ, ppm): 0.88 (t, 3H, -CH$_3$), 1.25 (m, 20H, -CH$_2$-), 1.72 (s, 6H, -CH$_3$), 1.91 (m, 2H, -CH$_2$-), 3.35 (t, 2H, -CH$_2$-), 4.19 (t, 2H, -CH$_2$-).

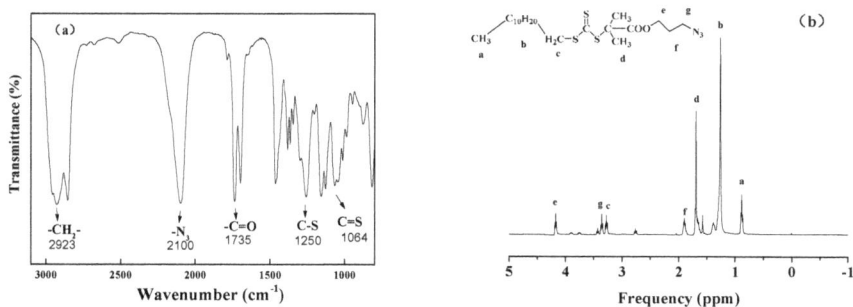

Figure 1. (a) FTIR spectra of the DMP-N$_3$; (b) ^{1}H spectra of the DMP-N$_3$.

For the typical polymerization processing of PMMA, 5 g of monomer MMA, chain transfer agent DMP-N$_3$ (0.168 g), and AIBN (8.2 mg), which were dissolved in dried anisole (20 ml), was added to a 50 ml Schlenk flask. The mixture solution was degassed with nitrogen using a three-way tube (three cycles). Then, a RAFT polymerization reaction was performed at 60 °C for 10 h. After polymerization, the azide-terminated PMMA was obtained when the unreacted MMA monomer was removed with THF for 24 h by Soxhlet apparatus. The resulting polymer was dried at 25 °C in a vacuum oven for 12 h. Scheme 1 shows the route of synthesis of azide-terminated PMMA. The resulting polymer was dried at 25 °C in a vacuum oven for 12 h. Scheme 1 shows the route of synthesis of azide-terminated PMMA.

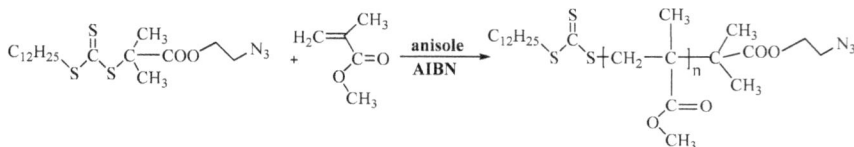

Scheme 1. The synthetic route for synthesis of azide-terminated poly(methyl methacrylate) PMMA via RAFT polymerization.

2.3. Alkyne-Modification of MWCNT (MWCNTs-alkyne)

In total, 2 g of the MWCNTs was subjected to acid treatment with 60 ml of an HNO$_3$:H$_2$SO$_4$ (1:3) mixture using ultrasound at 60 °C for 4 h, and then refluxed at 80 °C for 12 h. After the reaction, the treated MWCNTs (MWCNTs-COOH) were washed until the excess acid was completely removed.

In total, 1 g of MWCNTs-COOH was dispersed in thionyl chloride (65 mL) for 30 min. Then, 2 ml *N,N*-dimethylformamide (DMF) was added to this reaction mixture and stirred for 24 h at 70 °C. Thus, the MWCNT-COCl was obtained after being dried at 50 °C for 24 h. After that, 0.5 g MWCNT-COCl and 2 ml anhydrous triethylamine were mixed in 20 mL of trichloromethane. Following this, 3 ml of propargyl alcohol was added dropwise to the MWCNT-COCl mixture at 0 °C. The reaction between MWCNT-COCl and propargyl alcohol was carried out at room temperature for 10 h. Subsequently, the obtained MWCNTs-alkyne was purified by centrifugation and then dried at 50 °C in a vacuum oven for 24 h.

2.4. Preparation of MWCNTs-g-PMMA

In total, 0.1 g of MWCNTs-alkyne and 1 g of azide-terminated PMMA were mixed with 15 mL of DMF under ultrasonic treatment for 30 min. Then, a CuBr solution (0.0069 g dissolved 1 mL of water) was added. The Schlenk flask was degassed and back-filled with nitrogen, and then put in an oil bath at 50 °C. After the click reaction, MWCNTs-g-PMMA was purified by ethylenediaminetetraacetic acid (EDTA), THF, and ethanol centrifugation. Unreacted PMMA was removed with THF for 24 h by

Soxhlet apparatus and dried for 24 h in a vacuum oven at 50 °C. The synthesis steps from MWCNTs to MWCNTs-*g*-PMMA are shown in Scheme 2.

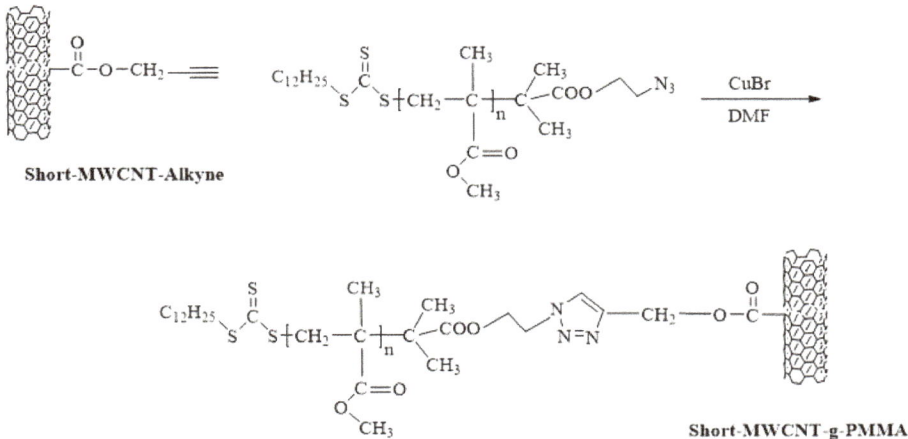

Scheme 2. Synthetic steps from multiwalled carbon nanotube (MWCNT) to MWCNT-*g*-PMMA.

2.5. Characterization

Fourier transform infrared spectroscopy (FTIR) spectra were performed on an Avatar 370 spectrometer (Nicolet, USA). The KBr pellet within an appropriate amount of MWCNTs was prepared. Raman spectra of the MWCNTs were gauged using a DXR Raman spectrometer (Thermo Scientific, USA) with the excitation wavelength of the laser at 532 nm. A laser intensity of 7.0 mW, an exposure time of 3 s, and the exposure rate of 20 times were applied in each measurement. An HP-6890 gas chromatograph (GC, Agilent, USA) measured the monomer conversion. Waters 515 gel permeation chromatography (GPC) was equipped with three columns (average pore sizes of 104, 105, and 106 nm, monodisperse polystyrene was used for the calibration standard sample). A Waters RI detector at 35 °C measured the molecular weight and molecular polydispersity of azide-terminated PMMA. Thermogravimetric analysis (TGA) was performed on a 209 F3 thermogravimetric (TG) analyzer (Netzsch Inc., Germany) under N_2 protection with a flow rate of 50 mL/min. The sample was heated from 50 °C to 700 °C at 10 °C/min. The grafting percentage (GP) for MWCNTs-*g*-PMMA can be calculated as shown in our previous work [24]. Transmission electron microscopy (TEM) was performed on a JEM-1200 EX/S transmission electron microscope (JEOL, Japan). Before observation, the dried MWCNTs were pretreated in THF under ultrasonic vibration for 20 min and then deposited on a covered copper grid.

3. Results and Discussions

3.1. RAFT Polymerization of Azide-Terminated PMMA

To investigate the efficiency of PMMA chain end transformation, polymerization kinetic studies on the linear PMMA synthesis were executed. The polymerization progress was checked by taking samples from the reaction mixture, which were measured by GPC and GC to estimate the evolution of conversion, molecular weights, and polydispersity index (PDI) with time. As it can be seen from Figure 2, M_n is increasing with conversion and PMMA has a small PDI, demonstrating a well-controlled RAFT reaction when using the RAFT agent. At the same time, when the high molecular weight is achieved at a high monomer conversion, the PDI values are still low (around 1.42). Compared with previous work [24], the conversion reaches a higher level (~55%), and the M_n of azide-terminated

PMMA increases more quickly with conversion, which suggests a much quicker reaction rate in this work.

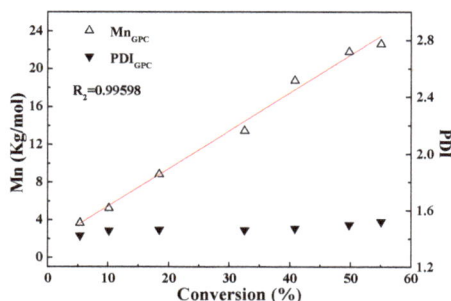

Figure 2. Dependence of number-average molecular weight (M_n) and PDI (M_w/M_n) of the grafted PMMA on the conversion for the PMMA RAFT polymerization at 60 °C with AIBN as initiator mediated by DMP-N_3 ([MMA]$_0$:[RAFT]$_0$:[AIBN]$_0$=250:1:0.2).

The results on the RAFT polymerization of the azide-terminated PMMA show a linear increase in $\ln(M_0/M_t)$ with time (Figure 3), suggesting a constant radical concentration, i.e., the absence of extensive termination reactions [25]. A conversion of 55% was obtained in 10 h, resulting in an azide-terminated PMMA with M_n = 22000 g/mol and PDI = 1.50.

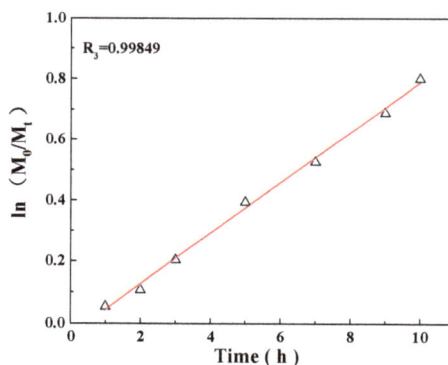

Figure 3. First-order kinetic plot for the RAFT polymerization of PMMA at 60 °C with AIBN as initiator mediated by DMP-N_3 ([MMA]$_0$:[RAFT]$_0$:[AIBN]$_0$=250:1:0.2).

In general, the MWCNTs-*g*-PMMA was obtained through the click coupling between MWCNTs-alkyne and azide-terminated PMMA. The grafting PMMA molecular weight was controlled by the prior synthesis of azide-terminated PMMA via RAFT polymerization. The grafting percentage on the MWCNT surface can present the click reaction efficiency with the reaction time. In this work, TGA measurement elucidated the grafting percentage that governed the alkyne/azide click reaction on MWCNT surfaces, as shown in Figure 4. It is evident that the grafting rate of PMMA on the surface of MWCNTs rises with the increase of reaction time. The grafting rate of MWCNTs-*g*-PMMA reaches 21.9% after a reaction of 24 h, and when the reaction continues to 30 h, the grafting rate is not changed obviously. The saturation grafting rate after 24 h could be due to the steric demands of the clicked PMMA chains rendering the remaining azido groups inaccessible for the click reaction [26].

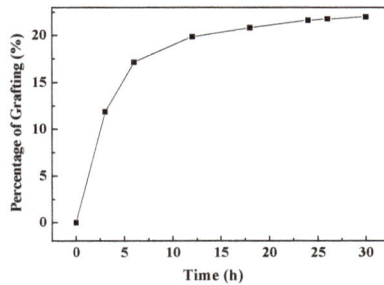

Figure 4. Grafting percentage of the azide/alkyne modified MWCNT.

3.2. TGA Analysis for the MWCNTS-g-PMMA

The TGA curves of pristine MWCNTs, MWCNTs-alkyne, and MWCNTs-*g*-PMMA are presented in Figure 5. Two stages of significant mass losses are observed in the curves of pristine MWCNTs. The first weight loss of about 0.5% happens before 200 °C, which is due to the loss of adsorbed water. The second step at a weight loss of approximately 1.4% is due to the impurities of pyrolysis. For the MWCNTs- alkyne, because of the decomposition of organic groups on the surface, the final weight loss was increased to 5.27%. For MWCNTs-*g*-PMMAs received at various polymerization times during the click reactions, because the grafted PMMA chains decompose to different extents, the final weight losses for the reaction time at 6 h, 12 h, and 18 h are 22.4%, 25.1%, and 26.9%, respectively (Figure 5c–e). The curves for the MWCNT-*g*-PMMA show two main decomposition steps at 205 °C and 405 °C, which correspond to the side chains and PMMA chains respectively. As measured, the grafting percentage (GP) for a reaction time of 6 h, 12 h, and 18 h is 17.2%, 19.9%, and 22.0%, respectively.

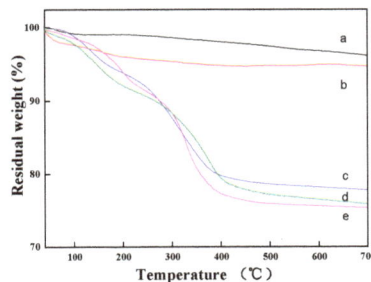

Figure 5. TGA results of pristine MWCNTs (**a**) MWCNTs-alkyne (**b**) MWCNTs-*g*-PMMA (**c–e**) polymerization for 6, 12, and 18 h, respectively.

3.3. Surface Structure Analysis for MWCNTS

The MWCNTs-alkyne was obtained via two gentle reaction processes, including acylchlorination and esterification, during which MWCNT-COOH was reacted with excess thionyl chloride to obtain a high esterification reaction efficiency COCL group, and the COCL group was then reacted with excess propargyl alcohol to get a complete surface coverage of the functional MWCNTs. FT-IR was performed on original and functionalized MWCNTs, and their corresponding spectra are shown in Figure 6. For the MWCNTs-COOH, the IR spectrum shows two absorption bands at 1740 cm^{-1} (corresponding to stretching vibrations of carbonyl groups C=O) and 1635 cm^{-1} (assigned to conjugated C=C stretching). MWCNTs-alkyne exhibits the same bands with the addition of an intensity band at 2100 cm^{-1} (the alkyne group) [27]. Azido-terminated PMMA presents a typical absorption band at 2100 cm^{-1} (-N$_3$ group), which suggests that the subsequent click reaction can be performed. After the click reaction, the new absorptions peak appeared at 1110 cm$^{-1,}$ and 1020 cm^{-1}, which was the C-O-C group in the ester group of the PMMA grafted onto the surface of MWCNTs. The click combination of the alkyne-functionalized MWCNTs and azide-functionalized PMMA provided a 1,2,3-triazole ring. This suggests that the PMMA molecule is successfully grafted onto the surface of the MWCNTs. Thus, the IR spectra of the MWCNTs-PMMA nanohybrid, featuring an alkyne peak of MWCNTs at 2100 cm^{-1}, entirely disappeared, revealing the formation of 1,2,3-triazole after the click reaction.

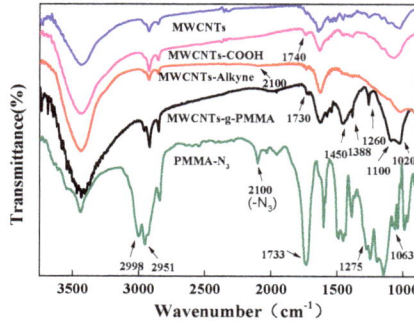

Figure 6. FT-IR spectra of pristine MWCNTs, MWCNTs-COOH, MWCNTs-alkyne, PMMA-N$_3$, and MWCNTs-*g*-PMMA (GP = 22.0%).

Figure 7 shows the Raman spectra of pristine MWCNTs, MWCNTs-COOH, MWCNTs-alkyne, and MWCNTs-*g*-PMMA. The characteristic peaks at 1343 cm^{-1} and 1585 cm^{-1} correspond to the D (tangential band) and G (disorder band) peaks of carbon nanotubes, respectively [22]. The D band is due to a disordered graphite structure or sp^3-hybridized carbons of the nanotubes, whereas the G band refers to a splitting of the E2g stretching mode of graphite [28,29]. The peak intensity D and G band ratios (I$_D$/I$_G$) for MWCNTs-COOH, MWCNTs-alkyne, and MWCNTs-*g*-PMMA are higher than those of the pristine MWCNTs, which suggests that alkyne-decoration and the click reaction successfully functionalize MWCNT.

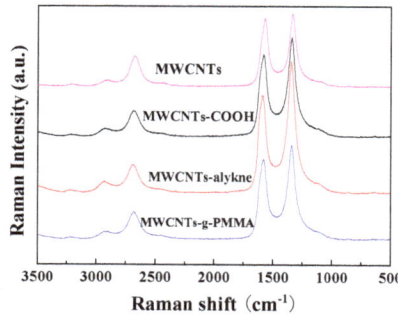

Figure 7. Raman spectra of pristine MWCNT, MWCNTs-COOH, MWCNTs-alkyne, and MWCNTs-g-PMMA (GP = 22.0%).

Figure 8 depicts the TEM photographs of pristine MWCNTs, carboxyl functionalized, and MWCNTs-*g*-PMMA. After the acid treatment, the closed end opening of the nanotube can be observed. The pristine MWCNTs generally present closed caps and cylindrical walls, which are uncapped and have rough "convex-concave" walls after the partial oxidation treatment [30]. These disfigurements improve the specific surface area and pore volume of the oxidized MWCNTs [31]. On the other hand, after the PMMA segments are grafted onto the surface of nanotubes by the click reaction, a thicker diameter of the MWCNTs-*g*-PMMA is observed.

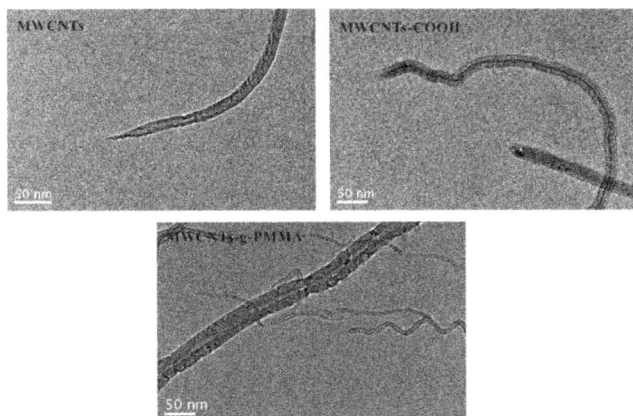

Figure 8. TEM photographs of the pristine MWCNTs, MWCNT-COOH, and MWCNTs-*g*-PMMA (GP = 22.0%).

Figure 9 shows the dispersion images of MWCNTs-*g*-PMMA in comparison with the pristine MWCNTs in water. After the ultrasonic treatment of these two dispersing solutions, the pristine MWCNTs could not disperse and stably suspended well in water because of the strong intrinsic van der Waals forces between them [13]. After two hours of standing, the pristine MWCNTs are obviously aggregated. However, the MWCNTs-*g*-PMMA can maintain its stable dispersion, even after 24 hours. This means that grafted PMMA has effectively reduced the apparent activation energy of MWCNTs, which can prevent the aggregation phenomenon. In future work, we will study this MWCNTs-*g*-PMMA dispersion in the polymeric membrane bulk.

Figure 9. The dispersion images of pristine MWCNTs (A) and MWCNTs-*g*-PMMA (GP = 22.0%) (B) in water. (**a**): after 0 h; (**b**): after 1 h; (**c**): after 2 h; (**d**): after 24 h.

4. Conclusions

In this work, to avoid the aggregation of MWCNTs, well-defined PMMA functionalized the combination of RAFT synthesized MWCNTs (MWCNTs-*g*-PMMA) and the clicked reaction. The success of PMMA grafting onto MWCNTs was determined by GPC, Raman spectroscopy, FT-IR, TGA, and TEM measurements. The kinetic reaction results show that the grafting percentage of PMMA chains grafted onto the MWCNTs surface rises with the increase of reaction time. Even at a low temperature (50 °C), the grafting rate of azide-terminated PMMA is comparatively fast during the click reaction when combining the alkyne-MWCNTs and azide-terminated PMMA in less than 24 h. As calculated by TGA analysis, the highest grafting degree of PMMA chains reaches 21.9%. Compared with the pristine MWCNTs, a thicker diameter of the MWCNTs-*g*-PMMA was observed by TEM, which confirmed that the PMMA chain grafted onto the surface of nanotubes. Therefore, the MWCNTs-*g*-PMMA could be dispersed and stably suspended in water, even for 24 h.

Author Contributions: W.M. conceived the experiments and wrote the manuscript; Y.Z. designed the experiments; Z.Z. and L.G. performed the click reaction experiments; Z.C. and Y.X. contributed to the results and discussions; H.Y. and F.G. gave the RAFT synthesis methods; J.Z. reviewed this manuscript and checked the English language corrections.

Funding: This research was funded by the Natural Science Foundation of China (Grant No. 21406017); the Natural Science Foundation of Jiangsu Province, China (Grant No. BK20140254); the Natural Science Foundation of the Jiangsu Higher Institutions of China (18KJA430005).

Acknowledgments: Thanks to the Priority Academic Program Development of Jiangsu Higher Education Institutions (PAPD) support.

Conflicts of Interest: The authors declare no conflict of interest.

References

1. Qu, X.L.; Alvarez, P.J.J.; Li, Q.L. Applications of nanotechnology in water and wastewater treatment. *Water Resour.* **2013**, *47*, 3931–3946. [CrossRef] [PubMed]
2. Bell, A.T. The impact of nanoscience on heterogeneous catalysis. *Science* **2003**, *299*, 1688–1691. [CrossRef] [PubMed]
3. Hummer, G.; Rasaiah, J.C.; Noworyta, J.P. Water conduction through the hydrophobic channel of a carbon nanotube. *Nature* **2001**, *414*, 188–190. [CrossRef] [PubMed]
4. Waghe, A.; Rasaiah, J.C.; Hummer, G. Filling and emptying kinetics of carbon nanotubes in water. *J. Chem. Phys.* **2002**, *117*, 10789–10795. [CrossRef]
5. Holt, J.K.; Park, H.G.; Wang, Y.M.; Stadermann, M.; Artyukhin, A.B.; Grigoropoulos, C.P.; Noy, A.; Bakajin, O. Fast mass transport through sub-2-nanometer carbon nanotubes. *Science* **2006**, *312*, 1034–1037. [CrossRef] [PubMed]
6. Majumder, M.; Chopra, N.; Andrews, R.; Hinds, B.J. Nanoscale hydrodynamics—Enhanced flow in carbon nanotubes. *Nature* **2005**, *438*, 44. [CrossRef]
7. Tunuguntla, R.H.; Henley, R.Y.; Yao, Y.C.; Pham, T.A.; Wanunu, M.; Noy, A. Enhanced water permeability and tunable ion selectivity in subnanometer carbon nanotube porins. *Science* **2017**, *357*, 792–796. [CrossRef]
8. Liu, Z.P.; Yang, R. Synergistically-enhanced thermal conductivity of shape-stabilized phase change materials by expanded graphite and carbon nanotube. *Appl. Sci.* **2017**, *7*, 574. [CrossRef]
9. Ma, W.; Gong, F.; Liu, C.; Tao, G.; Xu, J.; Jiang, B. SiO$_2$ reinforced HDPE hybrid materials obtained by the sol–gel method. *J. Appl. Polym. Sci.* **2014**, *131*, 596–602. [CrossRef]
10. Giovino, M.; Pribyl, J.; Benicewicz, B.; Kumar, S.; Schadler, L. Linear rheology of polymer nanocomposites with polymer-grafted nanoparticles. *Polymer* **2017**, *131*, 104–110. [CrossRef]
11. Fontananova, E.; Bahattab, M.A.; Aljlil, S.A.; Alowairdy, M.; Rinaldi, G.; Vuono, D.; Nagy, J.B.; Drioli, E.; Di Profio, G. From hydrophobic to hydrophilic polyvinylidenefluoride (PVDF) membranes by gaining new insight into material's properties. *RSC Adv.* **2015**, *5*, 56219–56231. [CrossRef]
12. Li, H.B.; Shi, W.Y.; Su, Y.H.; Zhang, H.X.; Qin, X.H. Preparation and characterization of carboxylated multiwalled carbon nanotube/polyamide composite nanofiltration membranes with improved performance. *J. Appl. Polym. Sci.* **2017**, *134*, e45268. [CrossRef]
13. Sanip, S.M.; Ismail, A.F.; Goh, P.S.; Soga, T.; Tanemura, M.; Yasuhiko, H. Gas separation properties of functionalized carbon nanotubes mixed matrix membranes. *Sep. Purif. Technol.* **2011**, *78*, 208–213. [CrossRef]
14. Balazs, A.C.; Emrick, T.; Russell, T.P. Nanoparticle polymer composites: Where two small worlds meet. *Science* **2006**, *314*, 1107–1110. [CrossRef] [PubMed]
15. Zdyrko, B.; Luzinov, I. Polymer brushes by the "grafting to" method. *Macromol. Rapid. Comm.* **2011**, *32*, 859–869. [CrossRef] [PubMed]
16. Ma, W.; Zhou, B.; Liu, T.; Zhang, J.; Wang, X. The supramolecular organization of PVDF lamellae formed in diphenyl ketone dilutions via thermally induced phase separation. *Colloid Polym. Sci.* **2013**, *291*, 981–992. [CrossRef]
17. Wu, P.; Feldman, A.K.; Nugent, A.K.; Hawker, C.J.; Scheel, A.; Voit, B.; Pyun, J.; Frechet, J.M.J.; Sharpless, K.B.; Fokin, V.V. Efficiency and fidelity in a click-chemistry route to triazole dendrimers by the copper(I)-catalyzed ligation of azides and alkynes. *Angew. Chem. Int. Ed.* **2004**, *43*, 3928–3932. [CrossRef] [PubMed]
18. Binder, W.H.; Sachsenhofer, R. 'Click' chemistry in polymer and materials science. *Macromol. Rapid Commun.* **2007**, *28*, 15–54. [CrossRef]

19. Escorihuela, J.; Marcelis, A.T.; Zuilhof, H. Metal-free click chemistry reactions on surfaces. *Adv. Mater. Interfaces* **2015**, *2*, 1500135.

20. Barner-Kowollik, C.; Du Prez, F.E.; Espeel, P.; Hawker, C.J.; Junkers, T.; Schlaad, H.; Camp, W.V. "Clicking" polymers or just efficient linking: What is the difference? *Ang. Chem. Int. Ed.* **2011**, *50*, 60–62. [CrossRef]

21. Zhao, L.J.; Zhao, F.Q.; Zeng, B.Z. Synthesis of water-compatible surface-imprinted polymer via click chemistry and RAFT precipitation polymerization for highly selective and sensitive electrochemical assay of fenitrothion. *Biosens. Bioelectron.* **2014**, *62*, 19–24. [CrossRef] [PubMed]

22. Pramanik, N.B.; Singha, N.K. Direct functionalization of multi-walled carbon nanotubes (MWCNTs) via grafting of poly(furfuryl methacrylate) using Diels-Alder "click chemistry" and its thermoreversibility. *RSC Adv.* **2015**, *5*, 94321–94327. [CrossRef]

23. Gondi, S.R.; Vogt, A.P.; Sumerlin, B.S. Versatile pathway to functional telechelics via RAFT polymerization and click chemistry. *Macromolecules* **2007**, *40*, 474–481. [CrossRef]

24. Ma, W.Z.; Zhao, Y.C.; Li, Y.X.; Zhang, P.; Cao, Z.; Yang, H.C.; Liu, C.L.; Tao, G.L.; Gong, F.H.; Matsuyama, H. Synthesis of hydrophilic carbon nanotubes by grafting poly(methyl methacrylate) via click reaction and its effect on poly(vinylidene fluoride)-carbon nanotube composite membrane properties. *Appl. Surf. Sci.* **2018**, *435*, 79–90. [CrossRef]

25. Can, A.; Altuntas, E.; Hoogenboom, R.; Schubert, U.S. Synthesis and MALDI-TOF-MS of PS-PMA and PMA-PS block copolymers. *Eur. Polym. J.* **2010**, *46*, 1932–1939. [CrossRef]

26. Chen, J.C.; Liu, M.Z.; Chen, C.; Gong, H.H.; Gao, C.M. Synthesis and characterization of silica nanoparticles with well-defined thermoresponsive PNIPAM via a combination of RAFT and click chemistry. *ACS Appl. Mater. Interfaces* **2011**, *3*, 3215–3223. [CrossRef] [PubMed]

27. Chang, Z.J.; Xu, Y.; Zhao, X.; Zhang, Q.H.; Chen, D.J. Grafting poly(methyl methacrylate) onto polyimide nanofibers via "click" reaction. *ACS Appl. Mater. Interfaces* **2009**, *1*, 2804–2811. [CrossRef] [PubMed]

28. Osswald, S.; Havel, M.; Gogotsi, Y. Monitoring oxidation of multiwalled carbon nanotubes by Raman spectroscopy. *J. Raman Spectrosc.* **2007**, *38*, 728–736. [CrossRef]

29. Jorio, A.; Dresselhaus, G.; Dresselhaus, M.S.; Souza, M.; Dantas, M.S.S.; Pimenta, M.A.; Rao, A.M.; Saito, R.; Liu, C.; Cheng, H.M. Polarized Raman study of single-wall semiconducting carbon nanotubes. *Phys. Rev. Lett.* **2000**, *85*, 2617–2620. [CrossRef]

30. Svrcek, V.; Pham-Huu, C.; Amadou, J.; Begin, D.; Ledoux, M.-J.; Le Normand, F.; Ersen, O.; Joulie, S. Filling and capping multiwall carbon nanotubes with silicon nanocrystals dispersed in SiO2-based spin on glass. *J. Appl. Phys.* **2006**, *99*, 064306. [CrossRef]

31. Hu, C.C.; Su, J.H.; Wen, T.C. Modification of multi-walled carbon nanotubes for electric double-layer capacitors: Tube opening and surface functionalization. *J. Phys. Chem. Solids* **2007**, *68*, 2353–2362. [CrossRef]

applied
sciences

MDPI

Article

The Effect of Functional Group Polarity in Palladium Immobilized Multiwalled Carbon Nanotube Catalysis: Application in Carbon–Carbon Coupling Reaction

Boggarapu Praphulla Chandra [1,2], Zheqiong Wu [3], Susana Addo Ntim [3], Golakoti Nageswara Rao [1] and Somenath Mitra [3,*]

[1] Department of Chemistry, Sri Sathya Sai Institute of Higher Learning, Prasanthi Nilayam 515134, India; bpc2sai@gmail.com (B.P.C.); nageswar.rao@rediffmail.com (G.N.R.)
[2] School of Science, Technology, Engineering and Mathematics, University of Washington Bothell, Bothell, WA 98011, USA
[3] Department of Chemistry and Environmental Science, New Jersey Institute of Technology, Newark, NJ 07102, USA; zw52@njit.edu (Z.W.); sa57@njit.edu (S.A.N.)
* Correspondence: somenath.mitra@njit.edu; Tel.: +1-973-596-5611; Fax: +1-973-596-3586

Received: 16 July 2018; Accepted: 29 August 2018; Published: 1 September 2018

Abstract: Carbon nanotubes (CNTs) are effective supports for nanometals and together they represent hybrids that combine the unique properties of both. A microwave-induced reaction was used to deposit nanopalladium on carboxylated and octadecylamine functionalized multiwall CNTs, which were used to carry out C–C coupling reactions in dimethylformamide (DMF) and toluene. These hybrids showed excellent catalytic activity with yield as high as 99.8%, while its enhancement with respect to commercially available Pd/C catalyst reached as high as 109%, and the reaction times were significantly lower. The polarity of the functionalized form was found to be a significant factor with the polar carboxylated CNT showing better activity in DMF while the relatively nonpolar octadecyl amine was better in toluene. The results suggest the possibility of tailor making functionalized CNTs when used as catalyst supports.

Keywords: Pd-CNT nanohybrids; functionalized CNTs; polarity; semi-homogeneous catalysis; heck reaction

1. Introduction

Hybrid structures involving carbon nanotubes (CNTs) and metal nanoparticles (NM) combine the unique properties of both [1]. These have been used in gas and liquid phase catalysis as well as electro catalysis [2–14]. Several CNT-NM hybrids have shown excellent reactivity [15–23], and different synthetic procedures have been used to immobilize Pt, Pd, Ru, and Rh NPs and a Rh/Pd alloy on CNTs. Various approaches [8–14] have been used to attach noble metals to CNTs [24–27]. These include in situ reduction [10], covalent bonding, electrostatic deposition, and physisorption [28]. Direct chemical reduction has been shown to increase metal loadings by as much as 50-fold [29] and microwave reactions have been reported as a fast method [30–33].

The pd-catalyzed reaction between aryl or alkenyl halides and olefins to deliver substituted olefins with high E-selectivity—known as the Heck reaction—is one of the most fundamental reactions in cross-coupling chemistry [34,35]. Initially, soluble Pd/Pt containing organometallic compounds were used as catalysts [34,35]. Later on, metal nanoparticle coated CNTs which represents the next generation of homogeneous catalyst were used to carry out these cross-coupling reactions. These have an advantage of easy separation of catalyst through centrifugation or precipitation [34–36].

In these typical reactions, the particle size, surface area, dispersion, and chemical nature of the support are important parameters [37–39]. In the case of CNT-NM, the presence of functional groups can alter many of these properties, but its role is not well understood. An added advantage of CNTs is that covalently introducing functional groups such as carboxylic carboxylic (–COOH), and (–CONH$_2$) can make them dispersible in different solvents and can be made solvent dispersible for different liquid phase reactions [40–42]. The synthesis of highly dispersed CNT-NM hybrids has been a challenge and is highly desirable for liquid phase reactions. We have reported the microwave induced synthesis CNT-Pt for catalytic hydrogenation of m-nitrochlorobenzene [19]. The objective of this work is to study the effect of CNT functionalization in CNT-Pd hybrids as applied to the Heck reaction and to study the effect of parameters such as dispersibility and polarity.

2. Experimental Section

2.1. Materials and Methods

Multiwall carbon nanotubes (MWCNT) (OD 20–30 nm, Purity 95%) were purchased from Cheap Tubes Inc. (Cambridgeport, VT, USA), and all other chemicals (iodobenzene, styrene, triphenylphosphine, (5 wt % Pd/C and PdCl$_2$)) were purchased from Sigma Aldrich (Bengaluru, Karnataka, India) with purity higher than 95%.

2.2. Synthesis of CNT-Metal Nanohybrids

Carboxylated Multiwall Carbon Nanotubes (MWCNT-COOH) were synthesized following a previously published methodology [42]. Briefly, the CNTs were functionalized in a Microwave Accelerated Reaction System (Mode: CEM Mars, CEM Corporation, 3100 Smith Farm Road, Matthews, NC, USA) fitted with internal temperature and pressure controls. Pre-weighed amounts of purified MWCNT were treated with a mixture of concentrated H$_2$SO$_4$ and HNO$_3$ solution under microwave radiation at 140 °C for 20 min. This led to the formation of carboxylic groups on the surface leading to high aqueous dispersibility. The resulting solid was filtered through a 10 μm membrane filter, washed with water to a neutral pH and dried under vacuum at 80 °C to a constant weight.

The MWCNT-COOH was used as the starting material to synthesize organic dispersible MWCNTs [43]. Pre-weighed amounts of MWCNT-COOH were mixed with thionylchloride (SOCl$_2$) and dimethylformamide (DMF) and were subjected to microwave radiation at 70 °C for 20 min leading to formation of MWCNT-COCl. The product was filtered and washed with THF till and was then dried in a vacuum oven at room temperature for 12 h. MWCNT-COCl was then reacted with octadecylamine (ODA) under microwave radiation at 120 °C for 10 min. The product was filtered and washed with hot ethanol and dichloromethane and dried at room temperature under vacuum to obtain MWCNT-ODA. The products MWCNT-COOH and MWCNT-ODA were used to synthesize the palladium loaded functionalized CNT nanohybrids: p-MWCNT-Pd and np-MWCNT-Pd respectively. The former was expected to be more polar. The p-MWCNT-Pd and np-MWCNT-Pd were synthesized by direct chemical reduction of metal salts as previously published by our group [19]. Pre-weighed amounts of the selected CNTs (MWCNT-COOH or MWCNT-ODA) were added to the reaction chamber together with a palladium dichloride (PdCl$_2$)-ethanol mixture. The reactions were carried out under microwave radiation at 190 °C for 10 min. After cooling, the products were filtered, washed with 0.5 N hydrochloric acid and MilliQ water to a neutral pH. The products (p-MWCNT-Pd or np-MWCNT-Pd) were dried at room temperature in a vacuum oven to constant weight.

The CNT-NM hybrids were characterized using a scanning electron microscope (SEM, Zeiss, Oberkochen, Germany) fitted with an Energy Dispersive X-ray spectrometer (EDX, Zeiss, Oberkochen, Germany), Thermogravimetric analysis (TGA, Perkin Elmer Inc., Waltham, MA, USA), and Fourier Transform Infrared spectroscopy (FTIR, Perkin Elmer Inc., Waltham, MA, USA). SEM data was collected on a LEO 1530 VP Scanning Electron Microscope equipped with an energy-dispersive X-ray analyzer. TGA was performed using a Pyris 1 TGA (Perkin-Elmer Inc., Waltham, MA, USA). FTIR measurements were carried out in purified Pottasium bromide (KBr) pellets using a Perkin-Elmer (Spectrum One) instrument (Perkin Elmer Inc., Waltham, MA, USA).

2.3. Catalytic C–C Coupling

Heck reaction was carried out as follows. CNT-NM (p-MWCNT-Pd or np-MWCNT-Pd) (3 mg, 0.01 mol %), iodobenzene (496.8 µL, 4.44 mmol), styrene (765.1 µL, 6.69 mmol), tributylamine (TBA) (2.014 mL, 8.92 mmol) and 0.5 mg PPh₃ were mixed in 5 mL of dry solvent in a two necked round bottom flask. The resulting suspension was placed in a preheated rota-mantle at 140 °C with magnetic stirring. The course of the reaction was followed by periodically withdrawing aliquots (50 µL) from the reaction mixture and analyzing by GC. The reaction was carried out with triphenylphosphine (PPh₃) in dimethylformamide (DMF) and toluene. The same reaction was carried out with commercial Pd/C (24 mg, 0.01 mol %) and also with PdCl₂ (2 mg, 0.01 mol %) to compare the catalytic activity of the CNT-NMs. The reaction is shown in Scheme 1.

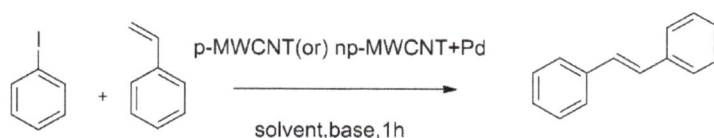

Scheme 1. General representation of reaction.

After 1 h reflux, the mixture was cooled for 15 min and extracted with diethyl ether. This was followed by washing with 2 N HCl and water to remove any base present. The ether layer was then dried over CaCl₂. Finally, the solution was filtered and dried in vacuum desiccator to constant weight. The obtained product was recrystallized from absolute alcohol and its composition was confirmed by comparison with pure stilbene. The final product did not show the presence of any byproducts by thin layer chromatography (TLC) analysis. The melting point of the recrystallized product and UV-λ_{max} were found to be 122.5 °C and 296 nm which was in line with what was expected for pure stilbene [44]. GC analysis of the product also showed only the presence of stilbene.

The aforementioned reactions were also carried out in microwave conditions at 420 W [45]. The amount of reactant, catalyst, and base (but no solvent) were maintained the same as done under reflux conditions. But, here the vessel was kept in microwave oven for 17 min where reaction went to completion with no reactant observed in the TLC analysis.

3. Results and Discussion

3.1. Nanohybrid Characterization

Figure 1 shows the SEM images of MWCNT-COOH, p-MWCNT-Pd, np-MWCNT-Pd and Pd/C. The MWCNTs had diameter in the range of 20–40 nm and the length was about 10–30 µm. There was no detectable change in tube morphology after acid treatment or after hybrid formation, implying that there was minimal visible damage to the tube structure. It is quite evident from the SEM images that the CNTs were coated with the metal nanoparticles. The EDX data shows large amounts of metal particles on the surface of the CNTs which are around 41.02% and 36.78% for p-MWCNT-Pd and np-MWCNT-Pd, respectively. The size of Pd particles ranged from 1 to 10 nm on p-MWCNT-Pd and from 5 to 50 nm on np-MWCNT-Pd. This implied that the catalyst was nanostructured and this was expected to contribute towards enhanced activity [19].

The FTIR spectra shown in Figure 2 confirmed the presence of functional groups in MWCNT-COOH, MWCNT-ODA, p-MWCNT-Pd, and np-MWCNT-Pd hybrids. The carbonyl stretching frequency in MWCNT-COOH was seen at 1716 cm⁻¹ (COOH). The 3440 cm⁻¹ band (O–H) present in MWCNT-COOH spectrum was attributed to the hydroxyl vibration of the carboxylic acid group introduced through functionalization. In Figure 2b, the peaks around 1636 cm^{-1} and 1715 cm^{-1} were attributed to stretching vibrations of the amide C=O group, the sharp peaks at 2921 cm^{-1} and 2847 cm^{-1} were attributed to the C–H stretching vibration of the alkyl chain from ODA, the peak at 3460 cm^{-1} was attributed to the N–H stretching vibration from ODA, and the peak at 1250 cm^{-1} was attributed to C–N stretching vibration. In all the samples, the peak around 1580 cm^{-1} was assigned to the C=C stretching of the carbon skeleton. The appearance of new bands at 993 cm^{-1} and 996 cm^{-1} in the spectra for the hybrids confirmed the presence of Pd [37–39]. The broad peak centered around 3440 cm^{-1} in MWCNT-COOH was not prominent in the IR spectrum of p-MWCNT-Pd, and the shift in the carbonyl stretching frequency from 1716 cm^{-1} to 1721 cm^{-1} in the spectra for the hybrids indicated interaction of Pd with the COOH group.

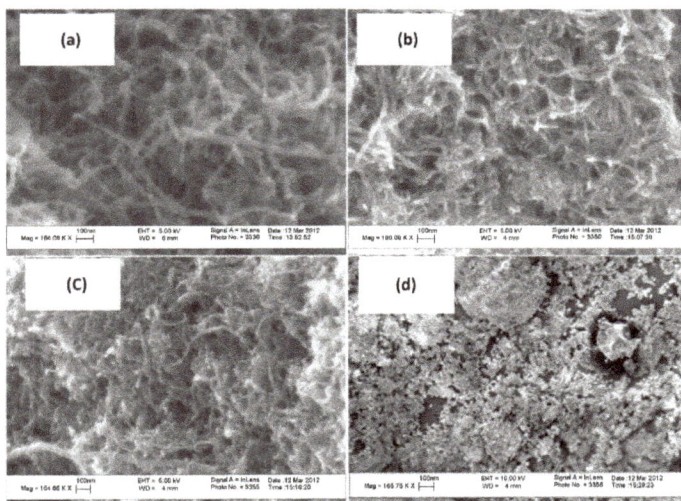

Figure 1. SEM images of (**a**) MWCNT-COOH, (**b**) p-MWCNT-Pd, (**c**) np-MWCNT-Pd, and (**d**) Pd/C.

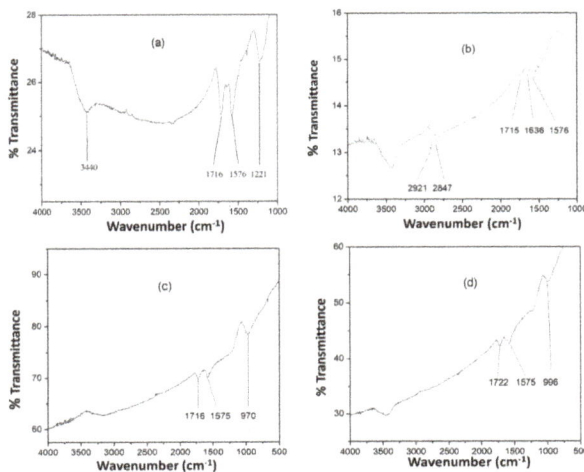

Figure 2. FTIR data for (**a**) MWCNT-COOH, (**b**) MWCNT-ODA, (**c**) p-MWCNT-Pd, and (**d**) np-MWCNT-Pd.

3.2. Dispersibility

Figure 3a–f shows the dispersibility of p-MWCNT-Pd (5 mg), np-MWCNT-Pd (5 mg), and Pd/C (5 mg) in *N,N*-dimethylformamide (5 mL), and toluene (5 mL). It is evident that p-MWCNT-Pd, np-MWCNT-Pd, and Pd/C were significantly more dispersible in DMF than in toluene. The particle size of the dispersed CNT-NM is presented in Table 1. It shows that there is not much difference between p-MWCNT-Pd and np-MWCNT-Pd in toluene, but there was significant difference in DMF. The particle size of CNT-NM agglomerates in toluene were higher than those in DMF. As can be seen in Figure 3, there was significant precipitation of the MWCNT-Pd in toluene, so the particle size represents what remained dispersed. The colloidal dispersion in DMF and TBA remained homogeneously suspended for several weeks without need for mechanical stirring.

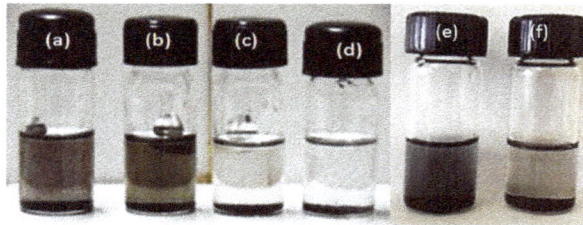

Figure 3. Photographs of different dispersions: (**a**) np-MWCNT-Pd dispersed in DMF, (**b**) p-MWCNT-Pd dispersed in DMF, (**c**) np-MWCNT-Pd in Toluene, (**d**) p-MWCNT-Pd in Toluene; (**e**) Pd/C dispersed in DMF, (**f**) Pd/C dispersed in Toluene.

Table 1. Particle size of dispersible CNT-NM.

Samples/Solvents	p-MWCNT-Pd	np-MWCNT-Pd
Toluene	240.3 nm	234.2 nm
Dimethylformamide (DMF)	168.2 nm	126.4 nm

The metal nanoparticles loading in the hybrid materials were quantified using TGA as shown in Figure 4. The resulting weight above 600 °C was attributed to the weight of residual metal. The nanohybrids were found to contain 43.02 and 38.93 percent by weight of Pd in p-MWCNT-Pd and np-MWCNT-Pd, respectively, while Pd/C contained 5% by weight of Pd.

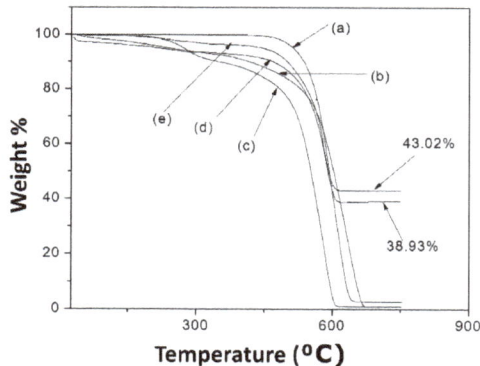

Figure 4. TGA of (**a**) MWCNT, (**b**) MWCNT-COOH, (**c**) MWCNT-ODA, (**d**) p-MWCNT-Pd, and (**e**) np-MWCNT-Pd.

3.3. Effect of CNT Functionalization

The CNT-NM hybrids were used as catalysts in the Heck reaction of iodobenzene with styrene. All the reactions were carried out keeping substrate to Pd molar ratio constant at 0.01 mole percent. In all cases, trans-stilbene was the only product detected which was established by the single spot in the TLC analysis and confirmed by GC analysis. This was also confirmed by the melting point of 122.5 °C and UV max at 296 nm of the isolated product.

The mechanism of the reaction in presence of PPh_3 is shown in Figure 5. In general, the Pd from the CNT-NM desorbed to a soluble form to form a loose complex such as $Pd(PPh_3)_4$. These complexes were more soluble than Pd and facilitated desorption of Pd from the catalyst. This may account for the high yield observed in the presence of PPh_3. At the end of the reaction, the Pd readsorbed on the CNT phase.

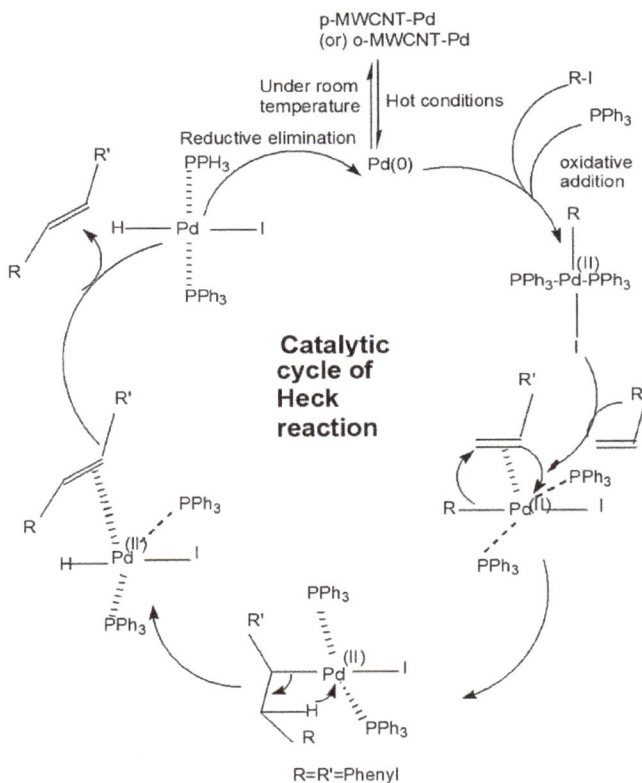

Figure 5. The role of CNT-Pd in the in the Heck reaction.

Table 2 presents isolated yields of all the reactions after the completion of the reaction and recrystallization. The results of reaction at 150 °C are presented in Table 2 and Figure 6. In general, shorter reaction times and greater yields were obtained with p-MWCNT-Pd and np-MWCNT-Pd compared to Pd/C, demonstrating the superiority of these nanohybrids in the reaction. The time to completion for the reaction for both p-MWCNT-Pd and np-MWCNT-Pd were 60 min in DMF compared to 90 min for Pd/C. Similar results are also seen in Figure 6, where yield enhancement of CNT-NM hybrids with respect to Pd/C is also presented.

Table 2. **Table 2.** Isolated yields of all the reactions at completion. Analysis was done after recrystallization.

S.No.	Catalyst	Reaction Completion Time (min)	% Yield	Yield Enhancement
Dimethylformamide (DMF) as solvent				
1	Pd/C	90	55.70	-
2	np-MWCNT-Pd	60	58.70	5%
3	p-MWCNT-Pd	60	99.78	79%
4	PdCl$_2$	120	45.67	-
Toluene as solvent				
5	Pd/C	105	10.90	-
6	p-MWCNT-Pd	90	14.39	32%
7	np-MWCNT-Pd	90	17.44	60%
Microwave (solvent free)				
8	p-MWCNT-Pd	17	90.01	-
9	np-MWCNT-Pd	17	66.92	-

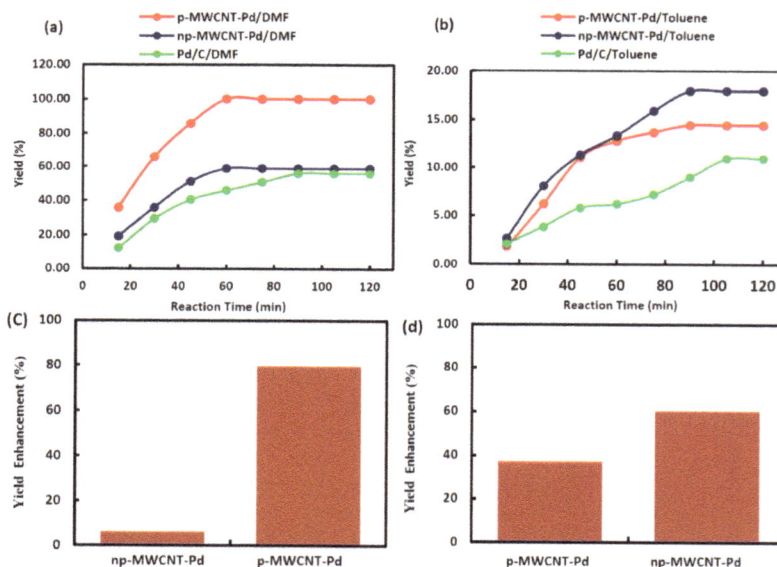

Figure 6. Yields of the CNT-metal nanohybrids and Pd/C catalyzed reactions (**a**) in DMF and (**b**) in toluene from GC analysis (Tri-butyl amine as base, temp 150 °C); (**c**) yield enhancement with respect to Pd/C in DMF; (**d**) enhancement with respect to Pd/C in toluene.

It is estimated that the ODA group will cause steric hindrance of the reactants. To test this hypothesis, the reactions were carried out under microwave conditions in absence of solvent. The reaction was complete in 17 min. During microwave heating, the CNT catalysts were seen to mix uniformly in the reaction mixture. The isolated yields for the reactions are shown in Table 2. The yield was much higher in the p-MWCNT-Pd than in np-MWCNT-Pd. The difference was attributed to steric effects, where the ODA group provided steric hindrances leading to lower yield.

In DMF, the p-MWCNT-Pd showed higher catalytic activity than np-MWCNT-Pd which was better than commercial Pd/C and PdCl$_2$. The yield for p-MWCNT-Pd reached as high as 99.78% in DMF, whereas np-MWCNT-Pd, Pd/C, and PdCl$_2$ showed yields of 58.7%, 55.7%, and 45.7%, respectively. The catalytic activity of p-MWCNT-Pd with larger particle size was higher than np-MWCNT-Pd with smaller particle size (Table 1). The p-MWCNT-Pd was more polar than np-MWCNT-Pd and its interactions with DMF, which is a polar solvent, were better. Also, the long C$_{18}$ chain from octadecyl group in np-MWCNT-Pd may have provided some steric hindrance to the reactants. Thus, the polarity and functionalized forms of support material are important in the matter of catalytic activity.

In the case of toluene as solvent, the results were the opposite. The np-MWCNT-Pd showed significantly higher catalytic activity than p-MWCNT-Pd, which was better than Pd/C. Compared to Pd/C, the enhancement for np-MWCNT-Pd and p-MWCNT-Pd were 60% and 32% respectively. Therefore, the enhancement nearly doubled for the nonpolar np-MWCNT-Pd. In this case, the catalytic activity of p-MWCNT-Pd with larger particle size was lower than np-MWCNT-Pd with smaller particle size. The steric hindrance from the octadecyl group did not reduce yield. Therefore, the higher yield of np-MWCNT-Pd was attributed to the higher interactions of the nonpolar np-MWCNT-Pd with nonpolar toluene. Figure 6a,b show yield as a function of time in DMF and toluene, respectively. As already mentioned, DMF was the better solvent with these catalysts, with yield reaching as high as 99.78% in 60 min. Yields were relatively lower and reaction times were longer in toluene, which was in line with what has been reported before [34,35]. In the case of DMF, the reactions were completed in 60 min with the nanohybrids and it took 90 min with Pd/C. On the other hand, when toluene was the solvent, the time taken for the same reactions were 90 min with the nanohybrids, and 105 min with Pd/C. Yield enhancement of CNT-NM hybrids with respect to Pd/C for the Heck reaction at 60 min reaction time in both DMF and toluene are shown in Figure 6c,d. In DMF, the enhancements were 79% and 5% for p-MWCN-Pd and np-MWCN-Pd, respectively, while the corresponding enhancements in toluene were 32% and 60% in toluene. It is evident that the polarities of the functionalized form of catalysts and solvents is the important factor for catalytic activity. DMF is a better solvent than toluene for Heck reaction due to the greater dispersibility of the catalyst in DMF and high boiling point. High boiling point of the solvent enables the reaction to be carried out at a relatively higher temperature. Also, for catalysts containing polar groups (–COOH in this case), a polar aprotic solvent like DMF is a suitable solvent for Heck reaction.

The reason for higher yields in DMF compared to the toluene (Boiling Point (B.P); 110.6 °C) is due to the better dispersibility of catalyst in addition to the higher boiling point of DMF (B.P; 153 °C). Nevertheless, when toluene was used as the solvent, np-MWCNT-Pd showed significantly higher catalytic activity than p-MWCNT-Pd, which was better than Pd/C. These were also attributed to dispersibility of catalyst (refer to Figure 6). Hence, it is evident that disparities in the polarity play an important role.

3.4. Active Metal Requirement and Reusability

Since DMF was the better solvent, the catalyst requirement for CNT-NM was calculated. Typical Heck reaction with conventional palladium salts requires 1–2 mol % catalyst [34,35]. In the present work with MWCNT-Pd hybrids, significantly lower quantities of Pd (order of 0.01 mol % Pd) resulted in higher yield at shorter reaction times. Therefore, this study showed that nan palladium on functionalized CNTs is more effective than conventional palladium salts. The turnover frequency (TOF) of the two catalysts were calculated as follows:

$$TOF = \frac{Product\ (mol)}{Pd\ (mol)\ \times\ Reaction\ time\ (h)} \tag{1}$$

Under reflux conditions, the TOF of p-MWCNT-Pd and np-MWCNT-Pd in DMF were 9977 and 5863, respectively, and in toluene 1090 and 977, respectively, which shows that the polar p-MWCNT-Pd in polar solvent has higher catalytic activity. Under microwave conditions, the TOF of p-MWCNT-Pd and o-MWCNT-Pd were 9000 and 6681, respectively.

An important issue has been the reusability of the catalyst. In the case of p-MWCNT-Pd and np-MWCNT-Pd, leaching of Pd from MWCNT support to hot DMF was demonstrated by carrying out a reaction under the aforementioned conditions and filtering the solid catalyst after 30 min while the mixture was still hot (around 150 °C). The filtrate was allowed to reflux in the absence of solid catalyst and yield from the reaction was determined. The yields were found to remain the same in the presence of the p-MWCNT-Pd as well as np-MWCNT-Pd. This demonstrated that Pd desorbed from the CNT support had remained suspended into the liquid phase in hot condition.

Irreversible leaching is a major consideration while dealing with solid catalysts due to depletion in the metal content and subsequent loss in activity. In another set of experiments in DMF with p-MWCNT-Pd and np-MWCNT-Pd, the recovered catalysts obtained by filtration after one hour reaction and subsequent cooling to room temperature were reused for a three consecutive reactions. The yield obtained with recovered catalysts in each step was within 2% of the previous step yield. This showed that there was no significant decrease in the catalytic activity of the recovered catalyst. This can be explained by "boomerang effect" [45–47] where catalyst which is dispersed into the bulk solution under the hot condition from the solid support redeposits under cold conditions. In order to study this, the reaction mixture (after 1-h reaction) was filtered in the hot condition to recover the catalyst. Similarly, the reaction mixture (after 1 h) was filtered after cooling to room temperature. The catalysts were analyzed for Pd content. The weight percent of Pd from the hot and cold conditions in the p-MWCNT-Pd were 11.9 and 22.9, respectively, as determined by EDX. This supported the existence of boomerang effect. In the future, poison testing will be conducted for these catalysts to understand the nature of catalysis which assist in designing next generation catalysts.

4. Conclusions

p-MWCNT-Pd and np-MWCNT-Pd were synthesized using a microwave induced reaction and they showed high catalytic activity compared to commercially available Pd/C. The CNT-Pd needed a smaller amount of active catalyst but led to faster reactions and higher yields. The best reaction conditions were found to be with DMF as solvent, TBA as base, and triphenylphosphine as ligand. While p-MWCNT-Pd showed better performance in polar DMF, np-MWCNT-Pd was more effective in toluene. Matching the polarity of CNT functionalization with that of the solvent is clearly an effective way to maximize yield.

Author Contributions: Conceptualization, S.M. and G.N.R.; Methodology, G.N.R. and B.P.C.; Software, Z.W. and S.A.N.; Validation, B.P.C., Z.W. and S.A.N.; Formal Analysis, B.P.C.; Investigation, B.P.C., Z.W. and S.A.N.; Resources, S.M. and G.N.R.; Data Curation, B.P.C., Z.W. and S.A.N.; Writing—Original Draft Preparation, B.P.C.; Writing—Review & Editing, B.P.C., S.M. and G.N.R.; Visualization, G.N.R. and B.P.C.; Supervision, S.M. and G.N.R.; Project Administration, S.M.; Funding Acquisition, S.M.

Funding: This work was funded by a grant from the National Institute of Environmental Health Sciences (NIEHS) under Grant No. R01ES023209. Partial support for this work was also provided by the Schlumberger Foundation Faculty for the Future Fellowship.

Acknowledgments: The authors are grateful to the founder Chancellor Bhagavan Sri Sathya Sai Baba, Sri Sathya Institute of Higher Learning, for his constant inspiration.

Conflicts of Interest: The authors of this manuscript certify that they have no conflict of interest regarding any financial interest, or non-financial in the subject matter or materials discussed in this manuscript. Any opinions, findings, and conclusions or recommendations expressed in this material are those of the authors and do not necessarily reflect the views of any funding agencies.

References

1. Yu, W.; Tu, W.; Liu, H. Synthesis of nanoscale platinum colloids by microwave dielectric heating. *Langmuir* **1999**, *15*, 6–9. [CrossRef]
2. Gong, K.; Chakrabarti, S.; Dai, L. Electrochemistry at carbon nanotube electrodes: Is the nanotube tip more active than the sidewall? *Angew. Chem. Int. Ed.* **2008**, *47*, 5446–5450. [CrossRef] [PubMed]

3. Ovejero, G.; Sotelo, J.; Rodríguez, A.; Díaz, C.; Sanz, R.; García, J. Platinum catalyst on multiwalled carbon nanotubes for the catalytic wet air oxidation of phenol. *Ind. Eng. Chem. Res.* **2007**, *46*, 6449–6455. [CrossRef]
4. Yang, X.; Cao, Y.; Yu, H.; Huang, H.; Wang, H.; Peng, F. Unravelling the radical transition during the carbon-catalyzed oxidation of cyclohexane by in situ electron paramagnetic resonance in the liquid phase. *Catal. Sci. Technol.* **2017**, *7*, 4431–4436. [CrossRef]
5. Iglesias, D.; Senokos, E.; Alemán, B.; Cabana, L.; Navío, C.; Marcilla, R.; Prato, M.; Vilatela, J.J.; Marchesan, S. Gas-phase functionalization of macroscopic carbon nanotube fiber assemblies: Reaction control, electrochemical properties, and use for flexible supercapacitors. *ACS Appl. Mater. Interfaces* **2018**, *10*, 5760–5770. [CrossRef] [PubMed]
6. Su, D.S.; Wen, G.; Wu, S.; Peng, F.; Schlögl, R. Carbocatalysis in Liquid-Phase Reactions. *Angew. Chem. Int. Ed.* **2017**, *56*, 936–964. [CrossRef] [PubMed]
7. Lamme, W.S.; Zečević, J.; de Jong, K.P. Influence of Metal Deposition and Activation Method on the Structure and Performance of Carbon Nanotube Supported Palladium Catalysts. *ChemCatChem* **2018**, *10*, 1552–1555. [CrossRef] [PubMed]
8. Kong, B.-S.; Jung, D.-H.; Oh, S.-K.; Han, C.-S.; Jung, H.-T. Single-walled carbon nanotube gold nanohybrids: Application in highly effective transparent and conductive films. *J. Phys. Chem. C* **2007**, *111*, 8377–8382. [CrossRef]
9. Kauffman, D.R.; Star, A. Chemically induced potential barriers at the carbon nanotube-metal nanoparticle interface. *Nano Lett.* **2007**, *7*, 1863–1868. [CrossRef] [PubMed]
10. Hu, X.; Wang, T.; Qu, X.; Dong, S. In situ synthesis and characterization of multiwalled carbon nanotube/Au nanoparticle composite materials. *J. Phys. Chem. B* **2006**, *110*, 853–857. [CrossRef] [PubMed]
11. Shi, Y.; Yang, R.; Yuet, P.K. Easy decoration of carbon nanotubes with well dispersed gold nanoparticles and the use of the material as an electrocatalyst. *Carbon* **2009**, *47*, 1146–1151. [CrossRef]
12. Hou, X.; Wang, L.; Zhou, F.; Wang, F. High-density attachment of gold nanoparticles on functionalized multiwalled carbon nanotubes using ion exchange. *Carbon* **2009**, *47*, 1209–1213. [CrossRef]
13. Tzitzios, V.; Georgakilas, V.; Oikonomou, E.; Karakassides, M.; Petridis, D. Synthesis and characterization of carbon nanotube/metal nanoparticle composites well dispersed in organic media. *Carbon* **2006**, *44*, 848–853. [CrossRef]
14. Zhang, M.; Su, L.; Mao, L. Surfactant functionalization of carbon nanotubes (CNTs) for layer-by-layer assembling of CNT multi-layer films and fabrication of gold nanoparticle/CNT nanohybrid. *Carbon* **2006**, *44*, 276–283. [CrossRef]
15. Yoon, B.; Wai, C.M. Microemulsion-Templated Synthesis of Carbon Nanotube-Supported Pd and Rh Nanoparticles for Catalytic Applications. *J. Am. Chem. Soc.* **2005**, *127*, 17174–17175. [CrossRef] [PubMed]
16. Planeix, J.; Coustel, N.; Coq, B.; Brotons, V.; Kumbhar, P.; Dutartre, R.; Geneste, P.; Bernier, P.; Ajayan, P. Application of carbon nanotubes as supports in heterogeneous catalysis. *J. Am. Chem. Soc.* **1994**, *116*, 7935–7936. [CrossRef]
17. Corma, A.; Garcia, H.; Leyva, A. Catalytic activity of palladium supported on single wall carbon nanotubes compared to palladium supported on activated carbon: Study of the Heck and Suzuki couplings, aerobic alcohol oxidation and selective hydrogenation. *J. Mol. Catal. A Chem.* **2005**, *230*, 97–105. [CrossRef]
18. Guo, D.-J.; Li, H.-L. High dispersion and electrocatalytic properties of palladium nanoparticles on single-walled carbon nanotubes. *J. Colloid Interface Sci.* **2005**, *286*, 274–279. [CrossRef] [PubMed]
19. Chen, Y.; Zhang, X.; Mitra, S. Solvent dispersible nanoplatinum-carbon nanotube hybrids for application in homogeneous catalysis. *Chem. Commun.* **2010**, *46*, 1652–1654. [CrossRef] [PubMed]
20. Mu, Y.; Liang, H.; Hu, J.; Jiang, L.; Wan, L. Controllable Pt nanoparticle deposition on carbon nanotubes as an anode catalyst for direct methanol fuel cells. *J. Phys. Chem. B* **2005**, *109*, 22212–22216. [CrossRef] [PubMed]
21. Zhang, K.; Zhang, F.J.; Chen, M.L.; Oh, W.C. Comparison of catalytic activities for photocatalytic and sonocatalytic degradation of methylene blue in present of anatase TiO_2–CNT catalysts. *Ultrason. Sonochem.* **2011**, *18*, 765–772. [CrossRef] [PubMed]
22. Chandrasekhar, P. CNT Applications in Microelectronics, Nanoelectronics, and Nanobioelectronics. In *Conducting Polymers, Fundamentals and Applications*; Springer: Berlin, Germany, 2018; pp. 65–72.
23. Karakoti, M. Surface Modification of Carbon-Based Nanomaterials for Polymer Nanocomposites. In *Carbon-Based Polymer Nanocomposites for Environmental and Energy Applications*; Ismail, A.F., Goh, P.S., Eds.; Elsevier: Amsterdam, The Netherlands, 2018; pp. 27–56.

24. Lordi, V.; Yao, N.; Wei, J. Method for supporting platinum on single-walled carbon nanotubes for a selective hydrogenation catalyst. *Chem. Mater.* **2001**, *13*, 733–737. [CrossRef]

25. Ellis, A.V.; Vijayamohanan, K.; Goswami, R.; Chakrapani, N.; Ramanathan, L.; Ajayan, P.M.; Ramanath, G. Hydrophobic anchoring of monolayer-protected gold nanoclusters to carbon nanotubes. *Nano Lett.* **2003**, *3*, 279–282. [CrossRef]

26. Giordano, R.; Serp, P.; Kalck, P.; Kihn, Y.; Schreiber, J.; Marhic, C.; Duvail, J.L. Preparation of rhodium catalysts supported on carbon nanotubes by a surface mediated organometallic reaction. *Eur. J. Inorg. Chem.* **2003**, *2003*, 610–617. [CrossRef]

27. Haremza, J.M.; Hahn, M.A.; Krauss, T.D.; Chen, S.; Calcines, J. Attachment of single CdSe nanocrystals to individual single-walled carbon nanotubes. *Nano Lett.* **2002**, *2*, 1253–1258. [CrossRef]

28. Azamian, B.R.; Coleman, K.S.; Davis, J.J.; Hanson, N.; Green, M.L. Directly observed covalent coupling of quantum dots to single-wall carbon nanotubes. *Chem. Commun.* **2002**, 366–367. [CrossRef]

29. Liu, Z.; Lin, X.; Lee, J.Y.; Zhang, W.; Han, M.; Gan, L.M. Preparation and characterization of platinum-based electrocatalysts on multiwalled carbon nanotubes for proton exchange membrane fuel cells. *Langmuir* **2002**, *18*, 4054–4060. [CrossRef]

30. Cano, M.; Benito, A.; Maser, W.K.; Urriolabeitia, E.P. One-step microwave synthesis of palladium–carbon nanotube hybrids with improved catalytic performance. *Carbon* **2011**, *49*, 652–658. [CrossRef]

31. Trusova, M.E.; Rodriguez-Zubiri, M.; Kutonova, K.V.; Jung, N.; Bräse, S.; Felpin, F.X.; Postnikov, P.S. Ultra-fast Suzuki and Heck reactions for the synthesis of styrenes and stilbenes using arenediazonium salts as super-electrophiles. *Org. Chem. Front.* **2018**, *5*, 41–45. [CrossRef]

32. Ichikawa, T.; Mizuno, M.; Ueda, S.; Ohneda, N.; Odajima, H.; Sawama, Y.; Monguchi, Y.; Sajiki, H. A practical method for heterogeneously-catalyzed Mizoroki–Heck reaction: Flow system with adjustment of microwave resonance as an energy source. *Tetrahedron* **2018**, *74*, 1810–1816. [CrossRef]

33. Díaz-Ortiz, Á.; Prieto, P.; Vázquez, E. Heck reactions under microwave irradiation in solvent-free conditions. *Synlett* **1997**, *3*, 269–270. [CrossRef]

34. Christoffel, F.; Ward, T.R. Palladium-catalyzed Heck cross-coupling reactions in water: A comprehensive review. *Catal. Lett.* **2018**, *148*, 489–511. [CrossRef]

35. Beletskaya, I.P.; Cheprakov, A.V. The Heck reaction as a sharpening stone of palladium catalysis. *Chem. Rev.* **2000**, *100*, 3009–3066. [CrossRef] [PubMed]

36. Yang, X.; Wang, X.; Qiu, J. Aerobic oxidation of alcohols over carbon nanotube-supported Ru catalysts assembled at the interfaces of emulsion droplets. *Appl. Catal. A* **2010**, *382*, 131–137. [CrossRef]

37. Daniel, M.-C.; Astruc, D. Gold nanoparticles: Assembly, supramolecular chemistry, quantum-size-related properties, and applications toward biology, catalysis, and nanotechnology. *Chem. Rev.* **2004**, *104*, 293–346. [CrossRef] [PubMed]

38. Moreno-Manas, M.; Pleixats, R. Formation of carbon-carbon bonds under catalysis by transition-metal nanoparticles. *Acc. Chem. Res.* **2003**, *36*, 638–643. [CrossRef] [PubMed]

39. Corain, B.; Kralik, M. Generating palladium nanoclusters inside functional cross-linked polymer frameworks. *J. Mol. Catal. A Chem.* **2001**, *173*, 99–115. [CrossRef]

40. Li, W.; Liang, C.; Qiu, J.; Zhou, W.; Han, H.; Wei, Z.; Sun, G.; Xin, Q. Carbon nanotubes as support for cathode catalyst of a direct methanol fuel cell. *Carbon* **2002**, *40*, 791–794. [CrossRef]

41. Li, W.; Liang, C.; Zhou, W.; Qiu, J.; Zhou, Z.; Sun, G.; Xin, Q. Preparation and characterization of multiwalled carbon nanotube-supported platinum for cathode catalysts of direct methanol fuel cells. *J. Phys. Chem. B* **2003**, *107*, 6292–6299. [CrossRef]

42. Chen, Y.; Mitra, S. Fast Microwave-Assisted Purification, Functionalization and Dispersion of Multi-Walled Carbon Nanotubes. *J. Nanosci. Nanotechnol.* **2008**, *8*, 5770–5775. [CrossRef] [PubMed]

43. Desai, C.; Addo Ntim, S.; Mitra, S. Antisolvent precipitation of hydrophobic functionalized multiwall carbon nanotubes in an aqueous environment. *J. Colloid Interface Sci.* **2012**, *368*, 115–120. [CrossRef] [PubMed]

44. Wheeler, O.H.; Pabon, H.N.B. Synthesis of Stilbenes. A Comparative Study1. *J. Org. Chem.* **1965**, *30*, 1473–1477. [CrossRef]

45. Sermon, P.; Bond, G. Hydrogen spillover. *Catal. Rev.* **1974**, *8*, 211–239. [CrossRef]

46. Sun, W.; Liu, Z.; Jiang, C.; Xue, Y.; Chu, W.; Zhao, X. Experimental and theoretical investigation on the interaction between palladium nanoparticles and functionalized carbon nanotubes for Heck synthesis. *Catal. Today* **2013**, *212*, 206–214. [CrossRef]

47. Bond, G.C. A Short History of Hydrogen Spillover. In *Studies in Surface Science and Catalysis*; Pajonk, G.M.S.J.T., Germain, J.E., Eds.; Elsevier: Amsterdam, The Netherlands, 1983; pp. 1–16.

Communication

Improvement in EMI Shielding Properties of Silicone Rubber/POE Blends Containing ILs Modified with Carbon Black and MWCNTs

Chao Liu, Chuyang Yu, Guolong Sang, Pei Xu * and Yunsheng Ding *

School of Chemistry and Chemical Engineering, and Anhui Key Laboratory of Advanced Functional Materials and Devices, Hefei University of Technology, Hefei 230009, China; f62889@163.com (C.L.); Ycy19990913@163.com (C.Y.); sglforget@163.com (G.S.)

* Correspondence: chxuper@hfut.edu.cn (P.X.); dingys@hfut.edu.cn (Y.D.); Tel.: +86-551-62901545 (Y.D.)

Received: 27 March 2019; Accepted: 22 April 2019; Published: 29 April 2019

Featured Application: The present study provides an ingenious and effective method to fabricate a promising and flexible composites for electromagnetic interference shielding applications.

Abstract: Silicone rubber (SR)/polyolefin elastomer (POE) blends containing ionic liquids modified with carbon blacks (CB-IL) and multi-walled carbon nanotubes (CNT-IL) were prepared by melt-blending and hot pressing. SR/POE/CB-IL and SR/POE/CB-CNT-IL composites showed co-continuous structural morphologies. The cation–π interactions between ILs and CNTs were stronger than those between ILs and CBs due to the large length and high surface area of CNTs, which promoted better dispersion of carbon fillers. SR/POE/CB-CNT-IL composites showed higher EMI SE than SR/POE/CB-IL composites containing identical filler contents because the CNTs with larger aspect ratios helped form more electrically-conductive networks.

Keywords: polymeric composites; silicone rubber; Ionic liquid; carbon materials; structural; EMI shielding

---

1. Introduction

In the rapidly-growing information age, silicone rubber has seen extensive application in the electronic and electrical equipment industries due to its excellent chemical resistance, non-toxic nature, heat-resistance, freeze-resistance, and good flexibility [1]. However, silicone rubber is electrically-insulating and transparent to electromagnetic radiation. As electronic information technology has developed and the use of electronic devices has become more widespread, electromagnetic interference (EMI), which interrupts the functionality of electronic devices and severely affects human organs, has become a serious problem [2–4]. Therefore, conductive rubber composites containing carbon materials, such as carbon blacks and multi-walled carbon nanotubes, have excellent electrical conductivities and flexibility and have been widely used as EMI shielding materials [5,6]. A typical approach is to add a single type of carbon material into a single kind of polymeric matrix to increase the shielding ability of the carbon-based polymer composites [7–10]. Ionic liquids (ILs), which consist of a pair of soft cationic and anionic species, are used to improve the dispersion of carbon fillers in the polymer matrix since ILs and carbon fillers exhibit strong cation–π physical interactions [11–14].

In this study, carbon blacks (CBs) and carbon nanotubes (CNTs) were modified with 1-vinyl-3-ethylimidazolium bromide ILs that contained carbon double bonds. The synergistic effect of ILs modified with CBs and CNTs on the morphology and the EMI shielding properties of silicone rubber/polyolefin elastomer/carbon black-carbon nanotube-ionic liquid (SR/POE/CB-CNT-IL) composites were systematically investigated to reveal the mechanism of the enhanced EMI shielding performance.

2. Experimental

Methyl vinyl silicone rubber (SR) with a vinyl content of 0.16 wt% was purchased from Wynca Co., Ltd., China. POE (grade 8150), with a melt flow index of 0.5 g/10 min at 190 °C/2.16 kg, was purchased from Dow Chemical Co., Ltd., USA. Conductive carbon blacks (CBs) with a BET surface area of 83 m^2/g were purchased from Evonik Degussa Co., Ltd., Germany. Multi-walled carbon nanotubes (CNTs, 10–20 nm in internal diameter and 10–30 μm in length) were supplied by Nanjing Xianfeng Nano Material Technology Co., Ltd., China. The 1-vinyl-3-ethylimidazolium bromide ILs were supplied by Lanzhou Greenchem ILs Co., Ltd., China. 2,5-dimethyl-2,5-di(tert-butylper-oxy)-hexane (DBPH) as curing agent was purchased from AkzoNobel Peroxide Co., China. EVA (ethylene-vinylacetate copolymer, grade 265, 28 wt% VA content) was purchased from Dupont Co., USA, and was used as a non-reactive compatibilizer to enhance the interfacial adhesion between SR and POE.

The SR/POE/CB-CNT-IL composites were prepared as follows. First, CBs, CBs, and CNTs, ILs-modified CBs (CB-IL), ILs-modified CBs and CNTs by (CB-CNT-IL) were ground in an agate mortar. Then, the fillers were added to SR by a two-roll mill at 125 °C, and POE and DBPH were introduced and mixed for 15 min. Finally, the composites were vulcanized under 10 MPa pressure for 10 min at 180 °C, and then post-cured in a drying oven at 190 °C for 3 h. The amounts of components prepared for the composites are shown in Table 1.

Table 1. Composite compositions (weight ratio).

Sample	SR	POE	CB	MWCNTs	ILs	EVA	DBPH
SR/POE	60	40	-	-	-	5	2
SR/POE/5CB-IL	60	40	5	-	1	5	2
SR/POE/10CB-IL	60	40	10	-	2	5	2
SR/POE/15CB-IL	60	40	15	-	3	5	2
SR/POE/20CB	60	40	20	-	-	5	2
SR/POE/20CB-IL	60	40	20	-	4	5	2
SR/POE/15CB-CNT-IL	60	40	10	5	3	5	2
SR/POE/20CB-CNT	60	40	15	5	-	5	2
SR/POE/20CB-CNT-IL	60	40	15	5	4	5	2

The morphology of the composites was observed by scanning electron microscopy (SEM, JEOL, JSM-6490LV, Japan) at an accelerating voltage of 20 kV and field-emission scanning electron microscopy (FESEM, JSM-6700F) at an accelerating voltage of 5 kV. The electrical conductivity was determined using a multi-function digital electric meter (Victor Tech, Victor 86-e). The EMI shielding effectiveness (SE) data were obtained by using nanocomposite slabs (22.86×10.16×0.6mm^3), which were fitted into a waveguide sample holder with a vector network analyser (Agilent Technologies, N5247A, America) at 8.0 to 12.0 GHz (X band). EMI shielding properties of the samples were evaluated from the S parameters, which could be calculated calculate SE$_T$ (total shielding effectiveness), SE$_R$ (reflection loss) and SE$_A$ (absorption loss) [15].

3. Results and Discussion

Figure 1 shows the SEM images of the fractured surfaces of SR/POE/CB-IL and SR/POE/CB-CNT-IL composites. The samples were first fractured in liquid nitrogen, and then the fractured surfaces were immersed into n-hexane at 90 °C for 12 h to remove the POE phase. All samples display typical co-continuous structures, and the collapsed surface morphology is attributed to the dissolution of the POE phase. Carbon fillers modified by ILs were premixed with SR and then mixed with POE. Most fillers were initially dispersed in the SR matrix because the localization of fillers is controlled by flow properties and dynamic factors. When CNTs were added, the viscosity of the SR phase increased significantly, and the volume fraction of the SR phase decreased significantly due to the large length and high surface area of CNTs. This caused the SR/POE/CB-CNT-IL composites to form regular and orderly co-continuous phase structures [16].

Figure 1. Scanning electron microscopy (SEM) images of the fracture surfaces of the silicone rubber/polyolefin elastomer/carbon black -ionic liquid (SR/POE/CB-IL) and silicone rubber/polyolefin elastomer/carbon black-carbon nanotube-ionic liquid (SR/POE/CB-CNT-IL) composites.

Figure 2 shows the FESEM images of the fracture surfaces of the SR/POE/CB-IL, SR/POE/CB, SR/POE/CB-CNT-IL and SR/POE/CB-CNT composites. When the ILs were added, the dispersion of carbon filler was improved significantly because the cation–π interactions between ILs and carbon fillers promoted the further dispersion of fillers and reduced agglomeration. In addition, when the CNTs were added, the cation–π interactions between ILs and CNTs were stronger than those between ILs and CBs due to the large length and high surface area of CNTs [17].

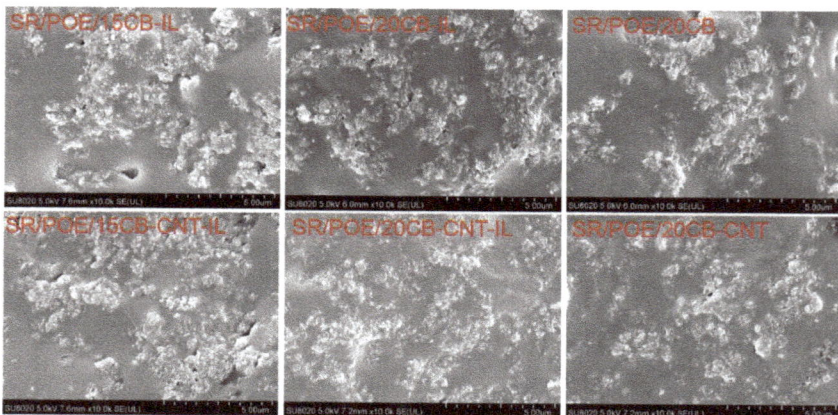

Figure 2. Field-emission scanning electron microscopy (FESEM) images of the fracture surfaces of the SR/POE/CB-IL, SR/POE/CB, SR/POE/CB-CNT-IL, and SR/POE/CB-CNT composites.

As the CB-IL content increased, the electrical conductivity of the SR/POE/15CB-IL and SR/POE/20CB-IL composites was 3.6×10^{-1} and 8.9×10^{-1} S/m, respectively. For the SR/POE/CB-CNT-IL composites, as the CB-CNT-IL content increased, the conductivity of the SR/POE/15CB-CNT-IL and SR/POE/20CB-CNT-IL composites was 6.7×10^{-1} and 3.5 S/m, respectively. CB-CNT-IL can regulate and control the aggregation, dispersion and uniformity of carbon fillers to form three-dimensional conductive networks. This improves the electrical conductivity, and the EMI SE values increased as the amount of fillers in the composites increased. Figure 3a shows the EMI shielding effectiveness (SE) of the composites in the frequency range of 8.0 to 12.0 GHz. SR/POE/CB-CNT-IL composites have higher EMI SE values than the SR/POE/CB-IL composites at identical filler loadings [18]. The CNTs with larger aspect ratios helped form a more conductive network and improved the EMI performance of the SR/POE/CB-CNT-IL composites. Specifically, the shielding effectiveness of the SR/POE/20CB-IL and SR/POE/20CB-CNT-IL composites was 29.4 dB and 36.5 dB, over the frequency range of 8.0 to 12.0 GHz, respectively. This gives the SR/POE/20CB-CNT-IL composite a novel potential electromagnetic interference that shields materials by eliminating electromagnetic pollution over a wide frequency range. To correlate the EMI shielding properties with the composite structures, both the average SE_T, SE_A, and SE_R values of SR/POE/CB-IL and SR/POE/CB-CNT-IL composites at 9.0 GHz were plotted with filler content as shown in Figure 3b. The SE_A and SE_R showed an increase as the filler content increased due to the synergistic effect between ILs and carbon fillers, which enhanced the electrical conductivity. Furthermore, the contribution of absorption to SE_T is greater than that of the reflection because network conductivity supports abundant interfaces which form multiple reflections and attenuate the incident EM waves [19]. The filler content in the SR/POE/CB-CNT-IL composites should be well above their percolation thresholds. Herein, the excellent EMI SE of the SR/POE/CB-CNT-IL composites is attributed to the presence of highly conductive networks, which effectively reflect, multi-scatter and adsorb the incident radiations. Figure 4a presents the SE_T, SE_A, and SE_R of SR/POE/CB, SR/POE/CB-CNT, SR/POE/CB-IL and SR/POE/CB-CNT-IL composites at 9.0 GHz. The EMI shielding performance of SR/POE/CB and SR/POE/CB-CNT composites obviously decreased compared with that of SR/POE/CB-IL and SR/POE/CB-CNT-IL composites. The results reveal that ILs promote the dispersion of carbon fillers due to the cation–π interactions between ILs and carbon fillers. Compared to SR/POE/CB-IL composites, the SR/POE/CB-CNT-IL composites with co-continuous structures have enhanced carrier mobility due to the presence of denser conductive networks and more interfacial structures formed by the uniformly-dispersed CB-CNT-IL. The schematic illustration of EMI shielding mechanisms for the SR/POE/CB-CNT-IL composites is shown in Figure 4b. The excellent EMI SE in this work can be attributed to the highly assembled carbon nano-fillers conductive networks that effectively reflect, multiple scatter and adsorb the incident radiation [20].

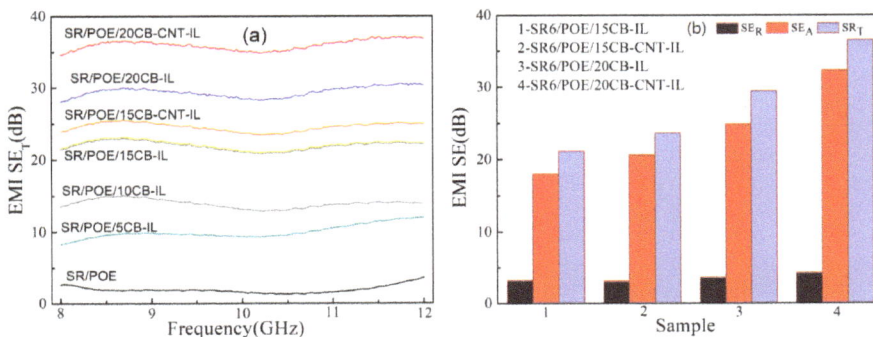

Figure 3. (a) Electromagnetic interference shielding effectiveness (EMI SE) of the SR/POE/CB-IL and SR/POE/CB-CNT-IL composites in the frequency range of 8.0 to 12.0 GHz, and (b) comparison of SE_T, SE_A, and SE_R of the SR/POE/CB-IL and SR/POE/CB-CNT-IL composites at 9.0 GHz.

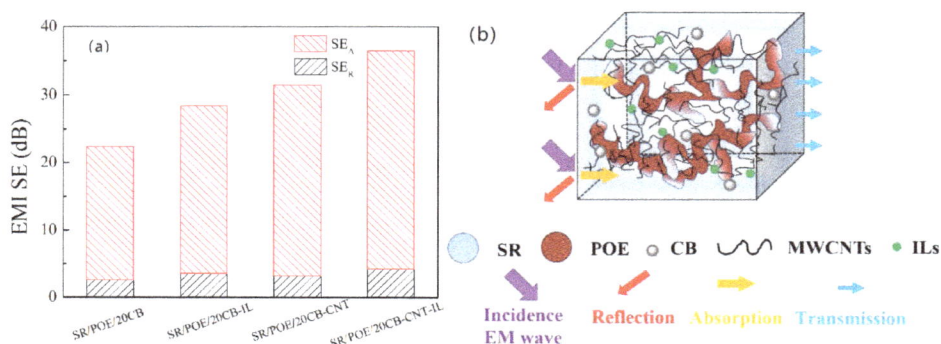

Figure 4. (**a**) Comparison of SE$_T$, SE$_A$, and SE$_R$ of the SR/POE/CB, SR/POE/CB-IL, SR/POE/CB-CNT-IL, and SR/POE/CB-CNT composites at 9.0 GHz, and (**b**) Schematic illustration of EMI shielding mechanisms for the SR/POE/CB-CNT-IL composites.

4. Conclusions

In summary, this study has investigated the morphology and EMI shielding properties of SR/POE/CB-CNT-IL composites, which were fabricated by a conventional two-roll mill mixing process. The cation–π interactions between ILs and CNTs were stronger than those between ILs and CBs due to the large length and surface areas of CNTs which promoted further dispersion of carbon fillers. The conductive networks play an important role in reflecting, scattering multiplication and absorption of the incident radiation. The SR/POE/CB-CNT-IL composites exhibited excellent EMI shielding properties due to the synergistic effect of ILs-modified CBs and CNTs. The EMI SE of the SR/POE/20CB-CNT-IL composite could reach 36.5 dB at 9.0 GHz. This gives the SR/POE/20CB-CNT-IL composite a novel potential electromagnetic interference that shields materials by eliminating electromagnetic pollution over a wide frequency range.

Author Contributions: Validation, G.S.; investigation, C.Y.; writing—original draft preparation, C.L.; writing—review and editing, P.X.; supervision and project administration, Y.D.

Funding: This research was funded by the Anhui Provincial Science and Technology Major Project (17030901074).

Conflicts of Interest: The authors declare no conflict of interest.

References

1. Li, Y.L.; Li, M.J.; Pang, M.L.; Feng, S.Y.; Zhang, J.; Zhang, C.Q. Effects of multi-walled carbon nanotube structures on the electrical and mechanical properties of silicone rubber filled with multi-walled carbon nanotubes. *J. Mater. Chem. C* **2015**, *3*, 5573–5579. [CrossRef]
2. Wang, H.B.; Teng, K.Y.; Chen, C.; Li, X.J.; Xu, Z.W.; Chen, L.; Fu, H.J.; Kuang, L.Y.; Ma, M.J.; Zhao, L.H. Conductivity and electromagnetic interference shielding of graphene-based architectures using MWCNTs as free radical scavenger in gamma-irradiation. *Mater. Lett.* **2017**, *186*, 78–81. [CrossRef]
3. Wang, G.L.; Wang, L.; Mark, L.H.; Shaayegan, V.; Wang, G.Z.; Li, H.P.; Zhao, G.Q.; Park, C.B. Ultralow-threshold and lightweight biodegradable porous PLA/MWCNT with segregated conductive networks for high-performance thermal insulation and electromagnetic interference shielding applications. *ACS Appl. Mater. Inter.* **2018**, *10*, 1195–1203. [CrossRef] [PubMed]
4. Engels, S.; Schneider, N.L.; Lefeldt, N.; Hein, C.M.; Zapka, M.; Michalik, A.; Elbers, D.; Kittel, A.; Hore, P.J.; Mouritsen, H. Anthropogenic electromagnetic noise disrupts magnetic compass orientation in a migratory bird. *Nature* **2014**, *509*, 353–356. [CrossRef] [PubMed]
5. Wang, R.; Yang, H.; Wang, J.L.; Li, F.X. The electromagnetic interference shielding of silicone rubber filled with nickel coated carbon fiber. *Polym. Test.* **2014**, *38*, 53–56. [CrossRef]

6.   Sudha, J.D.; Sivakala, S.; Kamlesh, P.; Nair, P.R. Development of electromagnetic shielding materials from the conductive blends of polystyrene polyaniline-clay nanocomposite. *Compos. Part A Appl. Sci. Manuf.* **2010**, *41*, 1647–1652. [CrossRef]

7.   Das, N.C.; Khastgir, D.; Chaki, T.K.; Chakraborty, A. Electromagnetic interference shielding effectiveness of carbon black and carbon fibre filled EVA and NR based composites. *Compos. Part A Appl. Sci. Manuf.* **2000**, *31*, 1069–1081. [CrossRef]

8.   Zeng, Z.H.; Jin, H.; Chen, M.J.; Li, W.W.; Zhou, L.C.; Zhang, Z. Lightweight and anisotropic porous MWCNT/WPU composites for ultrahigh performance electromagnetic interference shielding. *Adv. Funct. Mater.* **2016**, *26*, 303–310. [CrossRef]

9.   Wang, G.L.; Zhao, G.Q.; Wang, S.; Zhang, L.; Park, C.B. Injection-molded microcellular PLA/graphite nanocomposites with dramatically enhanced mechanical and electrical properties for ultra-efficient EMI shielding applications. *J. Mater. Chem. C* **2018**, *6*, 6847–6859. [CrossRef]

10.   Jin, L.; Zhao, X.M.; Xu, J.F.; Luo, Y.Y.; Chen, D.Q.; Chen, G.H. The synergistic effect of a graphene nanoplate/Fe$_3$O$_4$@BaTiO$_3$ hybrid and MWCNTs on enhancing broadband electromagnetic interference shielding performance. *RSC Adv.* **2018**, *8*, 2065. [CrossRef]

11.   Poothanari, M.A.; Abraham, J.; Kalarikkal, N.; Thomas, S. Excellent electromagnetic interference shielding and high electrical conductivity of compatibilized polycarbonate/polypropylene carbon nanotube blend nanocomposites. *Ind. Eng. Chem. Res.* **2018**, *57*, 4287–4297. [CrossRef]

12.   Kowsari, E.; Mohammadi, M. Synthesis of reduced and functional graphene oxide with magnetic ionic liquid and its application as an electromagnetic-absorbing coating. *Compos. Sci. Technol.* **2016**, *126*, 106–114. [CrossRef]

13.   Abraham, J.; Arif, M.P.; Kailas, L.; Kalarikkal, N.; George, S.C.; Thomas, S. Developing highly conducting and mechanically durable styrene butadiene rubber composites with tailored microstructural properties by a green approach using ionic liquid modified MWCNTs. *RSC Adv.* **2016**, *6*, 32493. [CrossRef]

14.   Abraham, J.; Arif, M.P.; Xavier, P.; Bose, S.; George, S.C.; Kalarikkal, N.; Thomas, S. Investigation into dielectric behaviour and electromagnetic interference shielding effectiveness of conducting styrene butadiene rubber composites containing ionic liquid modified MWCNT. *Polymer* **2017**, *112*, 102–115. [CrossRef]

15.   Li, Q.L.; Chen, L.; Ding, J.J.; Zhang, J.J.; Li, X.H.; Zheng, K.; Zhang, X.; Tian, X.Y. Open-cell phenolic carbon foam and electromagnetic interference shielding properties. *Carbon* **2016**, *104*, 90–105. [CrossRef]

16.   Cao, M.; Shu, J.J.; Chen, P.; Xia, R.; Yang, B.; Miao, J.B.; Qian, J.S. Orientation of boron nitride nanosheets in CM/EPDM co-continuous blends and their thermal conductive properties. *Polym. Test.* **2018**, *69*, 208–213. [CrossRef]

17.   Wang, P.; Xu, P.; Zhou, Y.Y.; Yang, Y.W.; Ding, Y.S. Effect of MWCNTs and P[MMA-IL] on the crystallization and dielectric behavior of PVDF composites. *Eur. Polym. J.* **2018**, *99*, 58–64. [CrossRef]

18.   Al-Saleh, M.H. Influence of conductive network structure on the EMI shielding and electrical percolation of carbon nanotube/polymer nanocomposites. *Synth. Met.* **2015**, *205*, 78–84. [CrossRef]

19.   Sang, G.L.; Dong, J.W.; He, X.T.; Jiang, J.C.; Li, J.B.; Xu, P.; Ding, Y.S. Electromagnetic interference shielding performance of polyurethane composites: A comparative study of GNs-IL/Fe$_3$O$_4$ and MWCNTs-IL/Fe$_3$O$_4$ hybrid fillers. *Compos. Part B Eng.* **2019**, *164*, 467–475. [CrossRef]

20.   Huang, H.D.; Liu, C.Y.; Zhou, D.; Jiang, X.; Zhong, G.J.; Yan, D.X.; Li, Z.M. Cellulose composite aerogel for highly efficient electromagnetic interference shielding. *J. Mater. Chem. A* **2015**, *3*, 4983–4991. [CrossRef]

Communication

Multi-Walled Carbon Nanotubes Composites for Microwave Absorbing Applications

Patrizia Savi [1,*], Mauro Giorcelli [2] and Simone Quaranta [3]

[1] Department of Electronics and Telecommunications (DET), Politecnico di Torino, 10129 Torino, Italy
[2] Department of Applied Science and Technologies (DISAT), Politecnico di Torino; 10129 Torino, Italy; mauro.giorcelli@polito.it
[3] Department of Information, Electronic and Telecom. Eng., La Sapienza University, 00184 Roma, Italy; quaranta@diet.uniroma1.it
* Correspondence: patrizia.savi@polito.it

Received: 27 December 2018; Accepted: 24 February 2019; Published: 27 February 2019

Featured Application: Potential application of the present research is in the field of radar invisible materials.

Abstract: The response of materials to impinging electromagnetic waves is mainly determined by their dielectric (complex permittivity) and magnetic (complex permeability). In particular, radar absorbing materials are characterized by high complex permittivity (and eventually large values of magnetic permeability), Indeed, energy dissipation by dielectric relaxation and carrier conduction are principally responsible for diminishing microwave radiation reflection and transmission in non-magnetic materials. Therefore, the scientific and technological community has been investigating lightweight composites with high dielectric permittivity in order to improve the microwave absorption (i.e., radar cross-section reduction) in structural materials for the aerospace industry. Multiwalled carbon nanotubes films and their composites with different kind of polymeric resins are regarded as promising materials for radar absorbing applications because of their high permittivity. Nanocomposites based on commercial multi-wall carbon nano-tube (MWCNT) fillers dispersed in an epoxy resin matrix were fabricated. The morphology of the filler was analyzed by Field emission scanning electron microscopy (FESEM) and Raman spectroscopy, while the complex permittivity and the radiation reflection coefficient of the composites was measured in the radio frequency range. The reflection coefficient of a single-layer structure backed by a metallic plate was simulated based on the measured permittivity. Simulation achievements were compared to the measured reflection coefficient. Besides, the influence of morphological MWCNT parameters (i.e., aspect ratio and specific surface area) on the reflection coefficient was evaluated. Results verify that relatively low weight percent of MWCNTs are suitable for microwave absorption applications when incorporated into polymer matrixes (i.e., epoxy resin).

Keywords: carbon nanotubes; composites; radar absorbing materials; complex permittivity

---

1. Introduction

An increased interest in carbon-based materials (graphene, single, and multiwalled carbon nanotubes) as reinforcements for polymers in order to improve specific material properties (from mechanical to electrical) has been emerging in the last few years [1–5]. Different kinds of composites based on a polymer matrix are present on the market. Specifically, composites based on the epoxy resin are used as high-performance materials because of their excellent mechanical properties, chemical resistance, thermal stability, and low production cost. Some examples of their applications are glues, adhesives, protective surface coatings, and electrical insulators [6–8]. Possible applications of these composites are also in microwave absorbers and EMI (electromagnetic interference) shielding applications [9–16]. Specifically, the tailoring of composites' complex dielectric permittivity is crucial for these two last applications. Indeed, high values of dielectric losses (ε''_r) stemming from both dielectric relaxation and conductivity ultimately lead to radiation absorption.

Thus, the maximization of dielectric losses is achieved by embedding particular fillers such as carbon nanotubes, graphene, or metallic nanoparticles into a polymer matrix.

In this paper, multi-wall carbon-nanotube (MWCNT) were dispersed in an epoxy resin polymer matrix in order to prepare a composite.. First, the morphology of the filler was investigated through field emission scanning electron microscopy (FESEM) and Raman spectra. The complex permittivity of pristine polymer matrix and composites with a given (4 wt%) was measured by means of a commercial dielectric probe. Experimental permittivity data were then used to compute the reflection coefficient of one-layer of a given thickness, backed by a metallic plate. Calculations were compared to the measured reflection coefficient of the composite. In addition, the effect of impurities, specific surface area and aspect ratio of the filler was also investigated. Experimental results confirmed the feasibility of epoxy resin-MWCNT (high aspect ratio) composites as microwave absorbers.

2. Materials and Methods

2.1. Materials Characteristics

MWCNTs (Nanothinx, Rio Patras, Greece) were selected for this work. Their characteristics, as declared by the manufacturer, are shown in Table 1.

Table 1. MWCNT characteristics as declared by the manufacturer.

Production method	CVD
Available form	Black powder
Diameter	6–15 nm
Length	$\geq 10~\mu m$
Number of layers	7–13
Carbon purity	93%
Metal particles (Al, Fe)	8 wt.%
Amorphous carbon	< 0.1%

MWCNTs were characterized by field emission scanning electron microscopy (FESEM, Zeiss Supra 40, (Oberkochen, Baden-Württemberg, Germany), energy-dispersive x-ray spectroscopy (EDX, Oxford Inca Energy 450, Oxford Instruments Abingdon-on-Thames, UK) and Raman spectroscopy (Renishaw Raman scope, 514.5 nm laser excitation, Wotton-under-Edge, United Kingdom). FESEM analysis was also used to investigate the MWCNTs dispersion in the polymer matrix (Epilox®, Leuna, Germany).

2.2. Sample Preparation and Characterization

Composite samples were prepared from a commercial bi-component thermosetting epoxy resin that acted as dispersing medium for carbon nanotubes. A commercial thermosetting resin (Epilox), produced by Leuna-Harze was used as a polymer matrix. It is a bi-component system formed by a resin and a hardener. Resin (T-19-36/700) is a commercial modified, colorless, low viscosity (650–750 mPa·s at 25 °C) epoxy resin with reduced crystallization tendency and a density of 1.14 g/cm^{-3}. The chemical composition of Epilox resin T19-36/700 is mainly Bisphenol A (30–60 wt.%), with the addition of crystalline silica (quartz) (1–10 wt.%), Glycidyl ether (1–10 wt.%), inner fillers (160 wt.%). Hardener (H10-31) is a liquid, colorless, low viscosity (400–600 mPa·s at 25 °C) modified cycloaliphatic polyamine epoxy adduct having density 1 g/cm^{-3}. A single 4 wt% carbon nanotubes loading was employed. 4 wt% choice was predicated on previous studies [17] concerning the dielectric properties of MWCNT/epoxy resin composites for different MWCNT amounts. Such an investigation had revealed that 4 wt% is a good trade-off between dispersion and dielectric behavior suitable for shielding applications. The sample preparation procedure is reported in [17,18]. Briefly, epoxy resin and MWCNTs were mixed by using Ultraturrax for 5 minutes. Hardener was added at the manufacturer-suggested ratio (1:2). Composite solution was poured into a silicon mold and degassed under vacuum in order to remove air bubbles during the polymerization process. Complete polymerization in air took 24 h.

The complex permittivity was measured by a commercial probe (Agilent 85070D) and a network analyzer (E8361A, Keysight Agilent, Santa Rosa, CA, U.S.). A standard calibration short/air/water was performed before each measurement. Several measurements were done on each sample and the average values reported.

2.3. Complex Permittivity Modeling

The complex permittivity of the prepared samples was modeled through a Maxwell-Garnett (MG) formulation [19]. Such a model can be applied to samples containing a filler volume fraction below the percolation threshold. The MG model was derived from a quasi-static approximation regarding the polarization of the whole sample.

According to the MG model, the effective relative permittivity of a multiphase mixture (isotropic, linear, and with inclusions small compared to the wavelength) can be calculated through the following equation [19,20]:

$$\varepsilon_{\text{eff}} = \varepsilon_b + \frac{\frac{1}{3}\sum_{i=1}^{n} f_i(\varepsilon_i - \varepsilon_b)\sum_{k=1}^{3}\frac{\varepsilon_b}{\varepsilon_b + N_{ik}(\varepsilon_i - \varepsilon_b)}}{1 - \frac{1}{3}\sum_{i=1}^{n} f_i(\varepsilon_i - \varepsilon_b)\sum_{k=1}^{3}\frac{N_{ik}}{\varepsilon_b + N_{ik}(\varepsilon_i - \varepsilon_b)}} \tag{1}$$

where ε_b is the relative permittivity of the polymer matrix (base dielectric), ε_i is the relative permittivity of the inclusions of type i, f_i is the fraction volume occupied by the inclusions of type i, N_{ik} is the depolarization factor of the inclusions of type i, k is each of the Cartesian coordinates (x, y, z), and n is the number of inclusion types.

The depolarization factors of canonical and oblate spheroids can be theoretically calculated. For instance, in the case of a needle. Like MWCNTs, the depolarization factors read [20]:

$$N_{i1} = 0, N_{i2} = N_{i3} = \frac{1}{2} \tag{2}$$

The electrical permittivity of the inclusions can be estimated from a theoretical model [19] by assuming that the conductive losses predominate over the dielectric relaxation:

$$\varepsilon_i(j\omega) = \varepsilon_i\prime - j\varepsilon_i{}'' = \varepsilon_i\prime - j\frac{\sigma_i}{\omega\varepsilon_0} \approx -j\frac{\sigma_i}{\omega\varepsilon_0} \tag{3}$$

where σ_i is the bulk conductivity of the i-th filler.

2.4. One-Layer Absorber Analysis and Reflection Coefficient Measurement Setup

An electromagnetic wave incident on a slab formed by a layer of composite, backed by a metallic plane (perfect electric conductor) as shown in Figure 1 (top) was considered. When the reflected wave has a half wave phase difference with respect to the incident field, there is destructive interference and the total field vanishes. The geometry of Figure 1 can be modeled considering a plane wave and a transmission line model (see Figure 1 bottom) to simulate the reflection coefficient, S_{11} [21]. Z_{air} and $Z_{\text{composite}}$ are the characteristic impedance of the equivalent lines of air and composites as defined in [12].

The reflection coefficient of small-sized backed by a metallic plane was measured in an anechoic chamber by means of a network analyzer (Rohde Schwarz, Munich, Germany) and two wide-band horn antennas. The specimen was positioned on a polystyrene cone pedestal. The distance between the pedestal and the two horn antennas was around 5 m. The floor around the pedestal was covered with standard pyramidal absorbers (see Figure 2). Time gating was performed in order to eliminate multiple reflections.

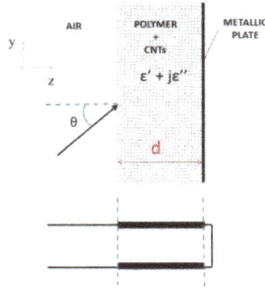

Figure 1. One-layer backed by a metallic plate (top) and transmission line model (bottom).

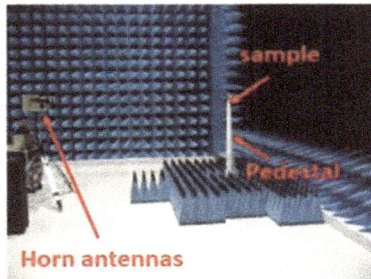

Figure 2. Reflection coefficient measurement setup.

3. Results

3.1. Filler Morphology Characterization

The FESEM images of MWCNTs are reported in Figure 3. In the inset, a high magnification (500 kX) picture shows that the diameters of MWCNTs are in the range 6–15 nm as stated by the manufacturer. Therefore, the high aspect ratio of the tubes tends to entangle them.

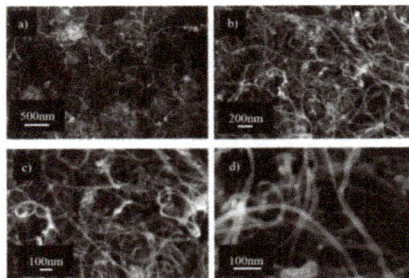

Figure 3. FESEM images of MWCNTs at different magnitudes: (**a**) 70 kX, (**b**) 130 kX, (**c**) 250 kX, (**d**) 500 kX.

EDX analysis results are reported in Figure 4. The composition of the MWCNTs is mainly carbon (C, 95.55 wt.%) with the presence of iron catalyst used for MWCNT growth (Fe, 3.52 wt.%). Traces of contaminants such as aluminum (Al, 0.83 wt.%) and sulfur (S, 0.10 wt.%) are also present.

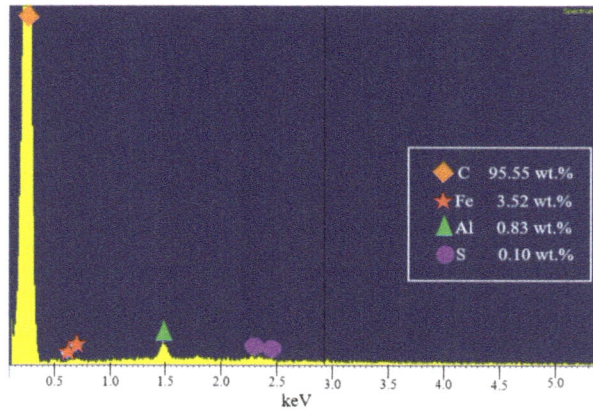

Figure 4. EDX spectra of MWCNTs.

Raman spectra and the associated relative curve fit are displayed in Figure 5. Usually, the Raman spectrum of the MWCNTs is divided into two main features. The first one, located in the 1000–1700 cm^{-1} range, is related to the G and D bands of carbon. In fact, they are the peaks used to estimate defects (D) and graphitization grade (G). The second range (1700–3500 cm^{-1}) is the second order of the Raman spectrum. D vibration overtone, G' or 2D band and the second order of D + G and 2G peaks, all belong to this wavenumber range. The calculated ratio (I_D/I_G = 0,95) shows a discrete organization of carbon structure with an intense D peak. The presence of 2D peak verifies the presence of defects [22].

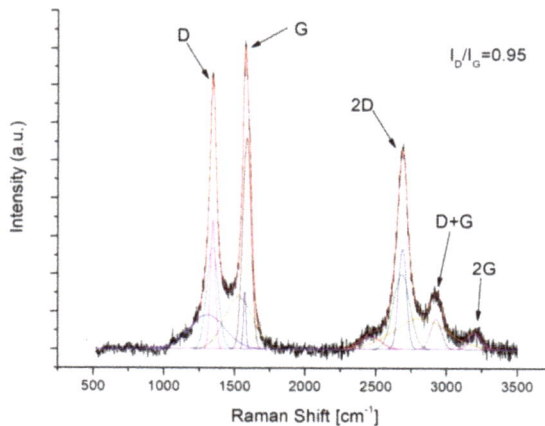

Figure 5. Raman spectra and its fit.

3.2. Composite Characterization

Composite samples were crio-fractured and analyzed at FESEM in order to evaluate the dispersion grade of the MWCNTs. Some significant images are reported in Figure 6. MWCNTs were not well dispersed into the polymer matrix due to the lack of any surface functionalization. Indeed, some agglomerates are still present like in the original material. Unfortunately, some agglomeration "comes with the territory" regardless of the dispersion method. Agglomerates reduce the electrical conductivity by decreasing the interconnection of conductive structures throughout the polymer matrix. Furthermore, agglomeration affects the dielectric properties of the material by lowering the amount of the filler surface area exposed to the binder. Consequently, the interfacial polarization decreases and both the real and imaginary parts of complex permittivity drop.

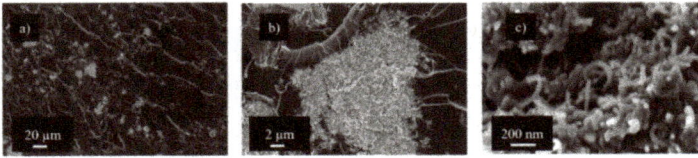

Figure 6. FESEM crio-fractured surface of the composite with different magnification (**a**) 1.0 kX, (**b**) 10.0 kX μm, (**c**) 200 kX

In Figure 7, the real part of the permittivity and the conductivity of the MWCNTs samples are shown together with the bare resin (Epilox). The measurements are compared to the Maxwell-Garnett model [19] as detailed in [23].

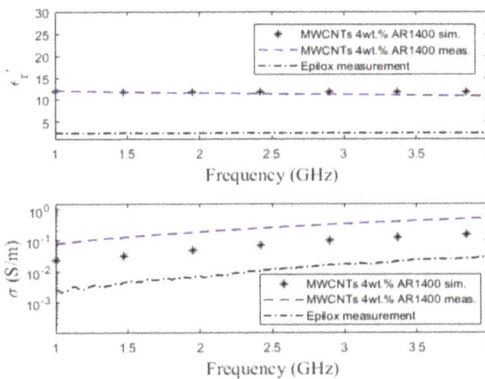

Figure 7. Real part of permittivity (top) and conductivity (bottom) versus frequency for 4wt% MWCNT-loaded epoxy resin: experimental measurements (dashed line) and simulations (stars); Pure epoxy (Epilox) resin (dash-dotted line).

As expected, both dielectric constant and the conductivity of the MWCNT-loaded resin are higher than the corresponding parameters of the bare Epilox. In fact, the introduction of carbon nanotubes into the polymer matrix results in the creation of an interfacial polarization component (ε'_r increase) and in the formation of a percolative conductive network across the composite (σ increase).

3.3. Analysis of One-Layer Absorber

Simulated results obtained considering a one-layer of thickness d = 2 mm and normal incidence are shown in Figure 8 (solid line) along with the measured data (dashed line) and the bare Epilox (dotted line). Practically no absorption occurs in the case of the resin alone. Obviously, the bare resin cannot benefit from any dielectric loss associated with either interfacial polarization or free-carrier conduction. On the other hand, the increase of the relative permittivity values upon the introduction of the MWCNTs into the polymer matrix produces a minimum in the reflection coefficient because of the increase in the interfacial polarization [24]. For a 4 wt.%, MWCNT concentration, a peak of −17 dB at 6 GHz is observed.

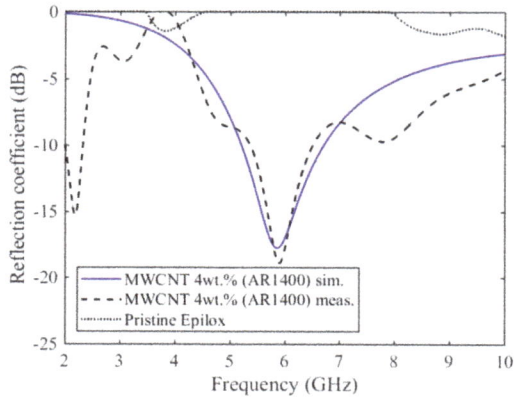

Figure 8. Microwaves reflection coefficient measurements: bare Epilox (dotted line) and 4 wt% MWCNT-loaded resin (dashed line). A comparison between the simulated (solid line) and the measured (black) reflection coefficient for the sample containing the carbon nanotubes is shown.

4. Discussion

As previously stated, microwave absorbing materials (i.e., materials that limit the reflection and transmission) require high dielectric losses (imaginary part of the complex permittivity) in order to convert the energy of the impinging radiation into heat and dissipate it. Needless to say, nanostructured materials such as CNTs have the capability to maximize both terms of the dielectric loss: relaxation and conductivity. A few deductions can be inferred from Figure 7 and 8. First, the high aspect ratio of the tubes (≈ 1400) in conjunction with their high specific surface area (≈ 300 m^2/g) brings about both a large interfacial polarization relaxing in the microwaves region and a good connectivity of the conductive structure across the composite [25]. Second, the presence of defects, as emerged from the Raman analysis, may lead to a further increase of the interfacial polarization. Hence, the relaxation of the interfacial polarization can account for the minimum in the reflection coefficient of the composite. Furthermore, a high aspect ratio and a high specific surface area can also cause multiple reflections (scattering), actually entrapping the radiation inside the material. Finally, both the high MWCNT aspect ratio and the relatively high (8 wt.%) content of metallic impurities enhance the composite conductivity and therefore the material losses.

In order to verify such hypotheses, a comparison of the results with a higher load (7 wt%) of structurally different MWCNTs was carried out. Figure 9 displays the real part of the complex permittivity and the conductivity for an Epilox resin loaded with 7 wt% of MWCNTs possessing the following characteristics: diameter (D) 22–45 nm, length (L) >10 μm, aspect ratio (AR = L/D) ≈ 300, catalyst residue 1.5 wt%, specific surface area (SSA) 200–250 m^2/g. For the sake of simplicity, this sample will be referred to as AR300 while the one employing 4 wt% of high aspect ratio (AR ≈ 1400) MWCNTs will be indicated as AR1400. Although the concentration of MWCNTs in the AR300 composite was almost doubled (i.e., compared to the AR1400), the both the dielectric constant and the conductivity dropped (for instance ε'_r (AR1400) = 12.1 and σ(AR1400) = 0.075 S/m; ε'_r (AR300) = 7.4 and σ (AR300) = 0.025 S/m at 1 GHz).

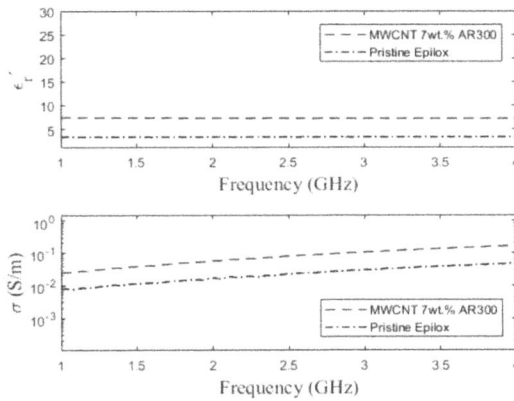

Figure 9. Dielectric constant and conductivity of a composite (i.e., Epilox matrix) containing 7 wt% of "low aspect ratio" (AR ≈ 300), high purity (98.5 wt%) MWCNTs.

Thus, the morphological properties of the carbon nanotubes proved to be crucial in determining the dielectric characteristic of the composite. In fact, the lower SSA and aspect ratio of the AR_{300} sample inevitably causes a reduction of the interfacial polarization. Consequently, the real component of the complex permittivity decreases. Apparently, also the conductivity seems to be affected by the smaller aspect ratio of the 7 wt% composite. However, the tube length is similar for both kinds of MWCNTs, and other parameters also need to be taken into account. For instance, the two samples differ in the wt% of the metal catalyst residue. Hence, the higher amount of metal particles in the AR1400 composite may also partially account for its higher conductivity. Such morphological (i.e., aspect ratio and specific surface area) and compositional (catalyst residue) features also reflect into the microwave absorption properties of the two materials. Figure 10 illustrates the reflection coefficient spectrum (for the AR300 composite. Despite the high concentration of carbon nanotubes compared to the AR1400 sample, the lower conductivity and dielectric constant resulted in similar amplitudes of the reflection coefficient.

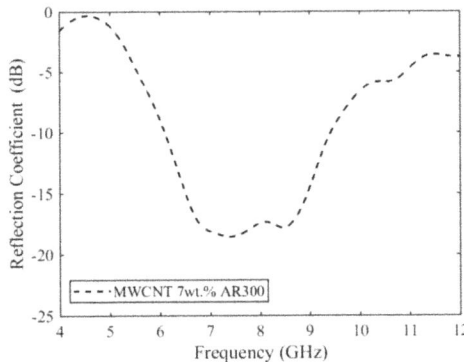

Figure 10. Reflection coefficient for the AR300 composite.

Thus, by optimizing the aspect ratio, specific surface area, and purity of the MWCNTs in order to achieve large interfacial polarizability and conductivity, it is possible to increase the microwave absorption of MWCNTs and design enhanced materials for EMI shielding.

Appl. Sci. **2019**, *9*, 851

5. Conclusions

Microwave absorbing nanocomposites based on the MWCNTs (4 wt%)/epoxy resin systems were prepared and characterized. The filler wt% was predicated on previous investigations exploring the relationship between MWCNT concentration, dispersibility, and complex permittivity. Indeed, 4 wt% represents a compromise between filler's uniform dispersion in the polymer matrix and high real and imaginary permittivity components. In fact, the formation of a conductive network throughout the composite volume allows for energy dissipation by both interfacial polarization and electron transport. Therefore, a 2 mm thick monolayer absorbing material was devised and tested for microwave reflection coefficient. The large interfacial polarization stemming from the MCWCNTs high aspect ratio (≈ 1400), and specific surface area ($300 \text{ m}^2/\text{g}$) resulted in a minimum in composite reflection coefficient at 6 GHz. Such an achievement is in good agreement with simulations. Moreover, dielectric and reflection coefficient measurements performed on epoxy resin composites comprised of morphologically-different MWCNTs demonstrated the aspect ratio, specific surface area, and purity represent dominant factors in determining the microwave absorption properties of the material (rather than the filler concentration).

Author Contributions: P.S. was responsible for permittivity and reflection coefficient measurements, Maxwell-Garnett model and microwave absorber model development and simulation, M.G. was responsible for the FESEM and Raman characterization, S.Q. was responsible for samples preparation; P.S. and S.Q. wrote the paper.

Funding: This research received no external funding

Acknowledgments: The authors would like to thank Nanothinx for supplying MWCNTs, Aamer Khan for help in composite preparation, Salvatore Guastella for FESEM and EDX analysis, and ing. Andrea Delogu, W. Ferrarese, and M. Mwanya (SELEX SE) for the reflection coefficient measurements.

Conflicts of Interest: The authors declare no conflict of interest.

References

1. Coleman, J.N.; Khan, U.; Blau, W.J.; Gun'ko, Y.K. Small but strong: A review of the mechanical properties of carbon nanotube–polymer composites. *Carbon* **2006**, *44*, 1624–1652. [CrossRef]
2. Bauhofer, W.; Kovacs, J.Z. A review and analysis of electrical percolation in carbon nanotube polymer composites. *Compos. Sci. Technol.* **2009**, *69*, 1486–1498. [CrossRef]
3. Park, J.H.; Alegaonkar, P.S.; Jeon, S.Y.; Yoo, J.B. Carbon nanotube composite: Dispersion routes and field emission parameters. *Compos. Sci. Technol.* **2008**, *68*, 753–759. [CrossRef]
4. Li, B.; Zhong, W.-H. Review on polymer/graphite nanoplatelet nanocomposites. *J. Mater. Sci.* **2011**, *46*, 5595–5614. [CrossRef]
5. Byrne, M.T.; Guin'Ko, Y.K. Recent advances in research on carbon nanotube–polymer composites. *Adv. Mater.* **2012**, *22*, 1672–1688. [CrossRef] [PubMed]
6. Xu, M.; Du, F.; Ganguli, S.; Roy, A.; Dai, L. Carbon nanotube dry adhesives with temperature-enhanced adhesion over a large temperature range. *Nat. Commun.* **2016**, *7*, 13450. [CrossRef] [PubMed]
7. Yu, S.; Tong, M.N.; Critchlow, G. Use of carbon nanotubes reinforced epoxy as adhesives to join aluminum plates. *Mater. Des.* **2010**, *31* (Suppl. 1), S126–S129. [CrossRef]
8. Sydlik, S.A.; Lee, J.H.; Walish, J.J.; Thomas, E.L.; Swager, T.M. Epoxy functionalized multi-walled carbon nanotubes for improved adhesives. *Carbon* **2013**, *59*, 109–120. [CrossRef]
9. Koledintseva, M.Y.; Drewniak, J.; DuBroff, R. Modeling of shielding composite materials and structures for microwave frequencies. *Prog. Electromagn. Res. B* **2009**, *15*, 197–215. [CrossRef]
10. Liu, L.; Kong, L.B.; Yin, W.-Y.; Matitsine, S. Characterization of Single- and Multi-walled Carbon Nanotube Composites for Electromagnetic Shielding and Tunable Applications. *IEEE Trans. Electromagn. Compat.* **2011**, *53*, 943–949. [CrossRef]
11. Micheli, D.; Pastore, R.; Apollo, C.; Marchetti, M.; Gradoni, G.; Primiani, V.M.; Moglie, F. Broadband electromagnetic absorbers using carbon nanostructure-based composites. *IEEE Trans. Microw. Theory Tech.* **2011**, *59*, 2633–2646. [CrossRef]

12. Savi, P.; Miscuglio, M.; Giorcelli, M.; Tagliaferro, A. Analysis of microwave absorbing properties of Epoxy MWCNT composites. *Prog. Electromagn. Res. Lett.* **2014**, *44*, 63–69. [CrossRef]
13. De Rosa, I.M.; Sarasini, F.; Sarto, M.S.; Tamburrano, A. EMC Impact of Advanced Carbon Fiber/Carbon Nanotube Reinforced Composites for Next-Generation Aerospace Applications. *IEEE Trans. Electromagn. Compat.* **2008**, *50*, 556–563. [CrossRef]
14. Saib, A.; Bednarz, L.; Daussin, R.; Bailly, C.; Lou, X.; Thomassin, J.M.; Pagnoulle, C.; Detrembleur, C.; Jerome, R.; Huynen, I. Carbon nanotube composites for broadband microwave absorbing materials. *IEEE Trans. Microw. Theory Tech.* **2010**, *1*, 2745–2754.
15. Zinenko, T.L.; Matsushima, A.; Nosich, A.I. Surface-plasmon, grating-mode and slab-mode resonances in THz wave scattering by a graphene strip grating embedded into a dielectric slab. Art. No. 4601809. *IEEE J. Sel. Topics Quant. Electron.* **2017**, *23*, 1–9. [CrossRef]
16. Khushnood, R.A.; Ahmad, S.; Savi, P.; Tulliani, J.-M.; Giorcelli, M.; Ferro, G.A. Improvement in electromagnetic interference shielding effectiveness of cement composites using carbonaceous nano/micro inerts. *Constr. Build. Mater.* **2015**, *85*, 208–216. [CrossRef]
17. Giorcelli, M.; Savi, P.; Yasir, M.; Yahya, M.H.; Tagliaferro, A. Investigation of Epoxy Resin/MWCNT composites behaviour at low frequency. *J. Mater. Res. Soft Nanomater. Focus Issue* **2015**, *30*, 101–107.
18. Giorcelli, M.; Savi, P.; Miscuglio, M.; Yahya, M.H.; Tagliaferro, A. Analysis of MWCNT/epoxy composites at microwave frequency: Reproducibility investigation. *Nanoscale Res. Lett.* **2014**, *9*, 1–5. [CrossRef] [PubMed]
19. Koledintseva, M.; DuBroff, R.E.; Schwartz, R.W. A Maxwell Garnett Model for Dielectric Mixtures Containing Conducting Particles at Optical Frequencies. *Prog. Electromagn. Res.* **2006**, *63*, 223–242. [CrossRef]
20. Sihvola, A.H.; Kong, J.A. Effective permittivity of dielectric mixtures. *IEEE Transactions Geosci. Remote Sens.* **1988**, *26*, 420–429. [CrossRef]
21. Ishimaru, A. *Electromagnetic Wave Propagation, Radiation, and Scattering*; Prentice Hall: Englewood Cliffs, NJ, USA, 1991.
22. Dresselhaus, M.S.; Dresselhaus, G.; Saitoc, R.; Joriod, A. Raman spectroscopy of carbon nanotubes. *Phys. Rep.* **2005**, *409*, 47–99. [CrossRef]
23. Mora, N.; Savi, P.; Giorcelli, M.; Rachidi, F. Analysis and Modeling of Epoxy/MWCNT Composites. In Proceedings of the International Conference on Electromagnetics in Advanced Applications (ICEAA2015), Torino, Italy, 7–11 September 2015.
24. Miscuglio, M.; Yahya, M.H.; Savi, P.; Giorcelli, M.; Tagliaferro, A. RF Characterization of polyer multi-walled carbon nanotube composites. In Proceedings of the IEEE Conference on Antenna measurements & Applications (CAMA), Antibes Juan-les-Pins, France, 16–19 November 2014; pp. 1–4.
25. Wang, H.; Zhu, D.; Zhou, W.; Fou, L. Electromagnetic and microwave absorbing properties of polyimide nanocomposites at elevated temperature. *J. Alloys Compd.* **2015**, *648*, 313–319. [CrossRef]

applied sciences

MDPI

Article

Fabrication of Novel CeO$_2$/GO/CNTs Ternary Nanocomposites with Enhanced Tribological Performance

Chunying Min [1,2,3,4,*], Zengbao He [1,2], Haojie Song [5], Dengdeng Liu [1,2], Wei Jia [1,2], Jiamin Qian [1,2], Yuhui Jin [1,2] and Li Guo [1,2,*]

[1] Research School of Polymeric Materials, Jiangsu University, Zhenjiang 212013, China; 2221705033@stmail.ujs.edu.cn (Z.H.); 2211605028@stmail.ujs.edu.cn (D.L.); 2211805067@stmail.ujs.edu.cn (W.J.); 3160705035@stmail.ujs.edu.cn (J.Q.); 3170705065@stmail.ujs.edu.cn (Y.J.)

[2] School of Material Science & Engineering, Jiangsu University, Zhenjiang 212013, China

[3] State Key Laboratory of Tribology, Tsinghua University, Beijing 100084, China

[4] National United Engineering Laboratory for Advanced Bearing Tribology, Henan University of Science and Technology, Luoyang 471023, China

[5] School of Materials Science & Engineering, Shaanxi University of Science & Technology, Xi'an 710021, China; songhaojie@sust.edu.cn

* Correspondence: mj790206@ujs.edu.cn (C.M.); liguo@ujs.edu.cn (L.G.); Tel.: +86-0511-8878-0190 (C.M.)

Received: 10 December 2018; Accepted: 29 December 2018; Published: 4 January 2019

Abstract: Increasing demands of multi-functional lubricant materials with well distributed nanoparticles has been generated in the field of oil lubrication. In this study, one-dimensional (1-D) acidified multi-walled carbon nanotubes (CNTs) and two-dimensional (2-D) graphene oxide (GO) sheets were dispersed together under an ultra-sonication condition to form CNTs/GO hybrids and the corresponding CNTs/GO hybrids decorated with uniform zero-dimensional (0-D) cerium oxide (CeO$_2$) nanoparticles were prepared via a facile hydrothermal method. The tribological performance of CeO$_2$/CNTs/GO ternary nanocomposite was systematically investigated using a MS-T3000 ball-on-disk tester. The results demonstrated that CeO$_2$/GO/CNTs nanocomposites can effectively reduce the friction of sliding pairs in paraffin oil. Moreover, the oil with 1 wt% of CeO$_2$/GO/CNTs exhibited the best lubrication properties with the lowest friction coefficient and wear scar diameters (WSD) compared with adding only GO nanosheet, CeO$_2$, and CeO$_2$/CNTs hybrid nanocomposite as lubricant additives. It is concluded that due to the synergistic effect of 0D CeO$_2$, 1D CNTs, and 2D GO during sliding process, a dimensionally mixed CeO$_2$/GO/CNTs nanocomposite exhibits excellent lubricating properties, providing innovative and effective additives for application in the field of lubrication.

Keywords: multi-walled carbon nanotubes; graphene oxide; cerium oxide; lubricating oil additives

1. Introduction

In the continuous development and research of tribology and nanomaterials, except for the improvement of traditional lubricant additives, many researchers focus on introducing nanomaterials into tribology for exploration [1–7]. Nanomaterials exhibit better anti-friction and wear resistance performance under more severe conditions, which makes them well-studied and applied in terms of friction [8,9]. Nanomaterials generally have a large specific surface area and surface tension, with the size between 1 and 100 nm. After the surface is modified, it can be uniformly and stably dispersed in some organic solvents [10–13]. The uniform dispersion of nanomaterials in the base oil can effectively improve its friction and wear resistance and possess excellent properties that other traditional lubricating oils do not have.

Cerium (Ce), a zero-dimensional (0D) nanoparticle, is a kind of rare earth element. Compared with metal nanoparticles, Ce is difficult to oxidize and exhibits excellent tribological properties in marine, aerospace, and other environments. Ce and its compounds as filler can effectively improve the tribological properties of the materials. Zhuo et al. [14] used CeO_2 and Cu nanoparticles as lubricating oil additives, and modified CeO_2 and Cu nanoparticles using a suitable surfactant mixture to achieve good dispersion and stability in the lubricating oil.

The special size, unique one-dimensional (1-D) tubular hollow and topology structure of carbon nanotubes make them attractive for various applications, including fabricating nanotubes reinforced nanocomposites on mechanical properties and tribological properties [15–17]. However, the insufficient dispersibility of carbon nanotubes deter their further research and development. The carbon nanotubes are acidified to have a partial oxygen-containing group on the surface to change their inertness. The modified carbon nanotubes are less likely to agglomerate and have increased solubility in solvent. Additionally, the modified carbon nanotubes are more easily combined with other materials during reaction. Gofman et al. [18] used carbon nanotubes containing a –COOH group as filling materials for polyimide, which showed significantly improved mechanical and tribological properties. Chen et al. [19] prepared carbon nanotubes using a mixture of sulfuric acid and nitric acid. The experimental results showed that the dispersion of the modified carbon nanotubes in base oil is improved, and the antifriction and wear resistance of the modified carbon nanotubes are also enhanced effectively.

Graphene oxide (GO), a two-dimensional (2-D) lattice, has numerous oxygen-containing groups and has superior thermal conductivity and mechanical properties. In addition, negative charge on the surface of GO makes it more convenient to combine with other materials without additional processing. As a new material, GO is widely used in various fields [20–22] such as tribology. Kim et al. [23] used an electrokinetic spraying process to coat graphene oxide on silicon wafer in order to use the coating as solid lubricant. Results showed that the coating has better surface protection ability and lower friction coefficient.

In recent years, many studies have confirmed that nano-hybrid materials can fully take advantage of various fillers and enhance the tribological properties of lubricating oil, polymers, and other materials [24–26]. Xin et al. [27] prepared a self-lubricating graphene oxide/nano-MoS_2 hybrid using a three-step method and discussed the synergistic effect of self-lubricating and anti-wear properties of the hybrid. Results show that the friction coefficient and wear rate of the base oil decreased by 25% and 64%, respectively, after the addition of nanocomposites. Zhang et al. [28] indicated that compared with pure liquid lubricant and liquid lubricant with MoS_2, the friction coefficient of liquid lubricant with CNTs/MoS_2 hybrids is decreased by 15%, due to the synergistic effect. Moreover, Bai et al. [29] studied a ceria/graphene composite as lubricating oil additive and considered the synergy between CeO_2 and graphene as an effective way to improve the tribological properties; however, the dispersion stability of lubricating additive and antifriction performance need further improvement.

In this work, we successfully prepared CNTs by using mixed acid. Additionally, different weight ratios of 0-D CeO_2 nanoparticle, 1-D CNTs, and 2-D GO were compounded to obtain various CeO_2/GO/CNTs nanocomposites via a hydrothermal method. CNTs and GO can form a stable structure through π-π bonding [30,31], and the surface is rich in oxygen-containing groups that can combine well with CeO_2. As far as we know, there are rare reports on the friction and wear properties of CeO_2/GO/CNTs hybrid as lubricating additive of base oil. Excellent structural properties of CeO_2/GO/CNTs nanocomposites can achieve synergistic effect among CeO_2, CNTs, and GO, which can provide some ideas for preparing high-performance oil lubricating additives.

2. Experimental

2.1. Materials

Cerium nitrate hexahydrate ($Ce(NO_3)_3 \cdot 6H_2O$), sodium hydroxide (NaOH), anhydrous ethanol were all purchased from Sinopharm Chemical Reagent Co., Ltd., Shanghai, China. Multi-walled carbon nanotubes (20–30 nm outer diameters, 10–30 μm length) were purchased from Chengdu organic chemistry Co., Ltd., Chengdu, China. GO was prepared using the sophisticated Hummers method [32] with natural flake graphite powders as raw material. Nitric acid (HNO_3) and sulfuric acid (H_2SO_4) were purchased from East Instrument Chemical Glass Co. Ltd., Zhenjiang, China.

2.2. Preparation of CeO_2/GO/CNTs Nanocomposites

Pure carbon nanotubes were treated with mixed acid of H_2SO_4 and HNO_3 (3:1) at 80 °C for 2 h under stirring, then washed and dried. The as-prepared CNTs and GO (total mass of GO and CNTs was 75 mg) were dispersed in a 37.5 mL mixed solution of ethanol and water using sonication to form the GO/CNTs uniform dispersion. Simultaneously, $Ce(NO_3)_3 \cdot 6H_2O$ was dispersed in a 37.5 mL mixed solution of ethanol and water, followed by pouring $Ce(NO_3)_3 \cdot 6H_2O$ solution into the GO/CNTs uniform dispersion with magnetic stirring for one hour. After that, the mixed solution was transferred into a Teflon-lined autoclave with addition of a certain amount of NaOH while gently stirring. The autoclave was thermal treated in an oven at 140 °C for 12 h. To obtain the CeO_2/GO/CNTs nanocomposites, the as-prepared products were washed by water and ethanol for five times, and freeze-drying for 24 h. CeO_2/CNTs nanocomposites was obtained following the above steps except that 75 mg GO/CNTs was substituted with 75 mg CNTs.

When the mass ratio of GO and CNTs was 3:1, GO/CNTs and $Ce(NO_3)_3 \cdot 6H_2O$ had different mass ratios (1:1, 3:1, 4:1) according to the same process to obtain various nanocomposites (CeO_2/GO/CNTs-a, CeO_2/GO/CNTs-b, CeO_2/GO/CNTs-c, respectively). When mass ratio of GO/CNTs and $Ce(NO_3)_3 \cdot 6H_2O$ was 3:1, GO/CNTs had different mass ratios (4:1, 5:1) according to the above steps to obtain various nanocomposites (CeO_2/GO/CNTs-d, CeO_2/GO/CNTs-e, respectively).

2.3. Characterization

The Raman measurements were performed on a Renishaw Microscope System RM2000 (Illinois, USA) with a 20 mW Ar^+ laser at 532 nm. The variation of elemental chemical state and surface chemical components were tested using X-ray photoelectron spectroscopy (XPS, ESCALAB 250XI, Thermo Fisher Scientific, Waltham, MA USA) with an Al Kα source. The morphologies and structures of different CeO_2/GO/CNTs nanocomposites were characterized using transmission electron microscopy (TEM, JEM-2100, Philips, Netherlands).

2.4. Analysis of Tribological Performance

The tribological properties of the specimen were determined by the friction test on the MS-T3001 ball-on-disk machine (the rotation speed was 300 r/min, the load was 10 N, the radius of rotation was 3 mm, and the duration was 30 min). The balls used in the experiment was made of GCr15 bearing steel with a hardness of 62 HRC and a diameter of 4 mm. The steel disk was made of 45 steel and polished to a bright surface with different specifications of sandpaper before the friction test. The width of the wear scar is an important parameter for characterizing the friction performance and measured using a Zeiss Observer Z1m metalloscope (Oberkochen Deutschland). The friction test was repeated three times for each specimen.

3. Results and Discussion

The flow diagram of the preparation of CeO_2/GO/CNTs nanocomposites is shown in Figure 1. The CNTs and GO nanosheets can combine together by covalent bonds due to the existence of a plurality of oxygen-containing groups on the surface of both CNTs and GO. In addition, Ce^{3+} can also be adsorbed on the surface of GO and CNTs by covalent bonding. Subsequently, NaOH solution was added to the mixed solution to provide a large amount of free OH^- which provided hydrolysis conditions for Ce^{3+}. The hydrothermal reaction is carried out to form $Ce(OH)^{2+}$, $Ce(OH)_2^+$ or $Ce(OH)_3$ compound, which were converted into CeO_2 via a hydrothermal reaction under high temperature and pressure. Then, in the reaction vessel, these products are converted into more stable CeO_2 with the high temperature and pressure hydrothermal reaction progresses, and finally the $CeO_2/GO/CNTs$ nanocomposite is obtained.

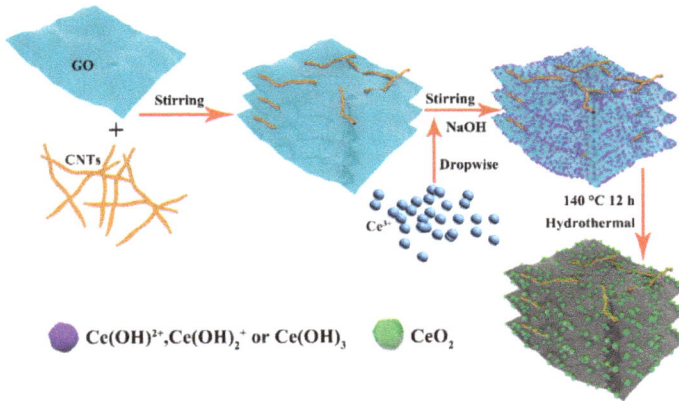

Figure 1. The formation processes of $CeO_2/GO/CNTs$ nanocomposites.

The morphologies of $CeO_2/CNTs$ and $CeO_2/GO/CNTs$ nanocomposites were characterized by TEM as shown in Figure 2. In Figure 2a, numerous CeO_2 nanoparticles were unevenly deposited on CNTs which were intertwined with each other accompanied by agglomeration. Figure 2b–f shows the microstructure of $CeO_2/GO/CNTs$ with different weight ratios of CeO_2, GO, and CNTs. From Figure 2b–f, it can be seen that GO exists as a carrier where CNTs combines and CeO_2 nanoparticles deposits. Specifically, via TEM observation of $CeO_2/GO/CNTs$-d nanocomposites (Figure 2e), CNTs and CeO_2 nanoparticles displayed excellent uniformity compared with other nanocomposites. It can be speculated that $CeO_2/GO/CNTs$-d nanocomposite is more suitable as lubricant additives in terms of the micromorphology of nanocomposites.

Raman spectroscopy is an efficient approach to prove the electronic structure of $CeO_2/GO/CNTs$ nanocomposites, especially for the changes of the C=C, C–C, and defects. Figure 3 shows the Raman spectra of GO, $CeO_2/CNTs$, and $CeO_2/GO/CNTs$ nanocomposites. From the Raman spectra of GO, there are characteristic bands related to the D ($\approx$1333 cm^{-1}) and G (1583 cm^{-1}) band in both GO and CNTs [33,34]. Apart from the D and G bands, the distinct strong peak at 456.13 cm^{-1} corresponds to the characteristic peak of CeO_2 nanoparticles [35]. However, in the Raman spectra of the $CeO_2/GO/CNTs$ nanocomposite, there were only one set of characteristic peaks, which appeared at 1333.54 cm^{-1} and 1583.53 cm^{-1} due to the coincidence of the D and G peaks of GO and CNTs. Ordinarily, the degree of the defects of the composites is characterized by the ratio of the D peak and G peak (I_D/I_G) [36–38]. In comparison with the I_D/I_G of GO and $CeO_2/CNTs$ (0.94, 0.99), the I_D/I_G of $CeO_2/GO/CNTs$-d nanocomposite (1.05) increased. These results implied the changes in the structure of surface functional groups of nanocomposites and the formation of the $CeO_2/GO/CNTs$ nanocomposite. Specific results can also be observed in the XPS survey spectrum (Figure 4).

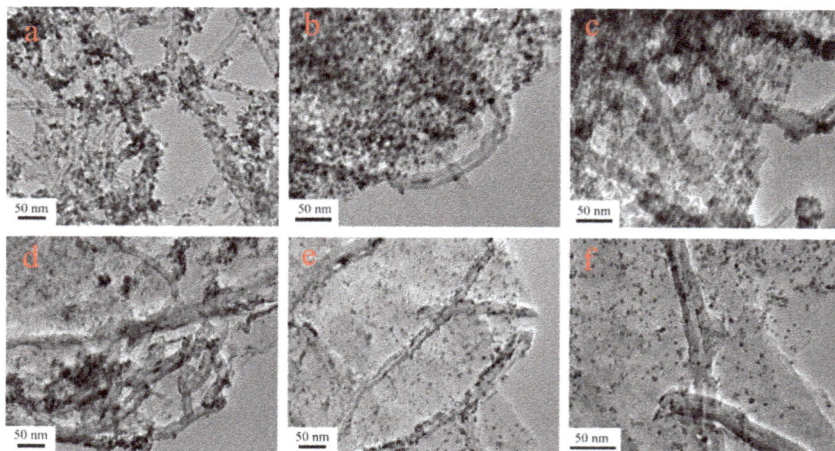

Figure 2. TEM images of (**a**) CeO_2/CNTs nanocomposites; (**b**) CeO_2/GO/CNTs-a nanocomposites; (**c**) CeO_2/GO/CNTs-b nanocomposites; (**d**) CeO_2/GO/CNTs-c nanocomposites; (**e**) CeO_2/GO/CNTs-d nanocomposites; and (**f**) CeO_2/GO/CNTs-e nanocomposites.

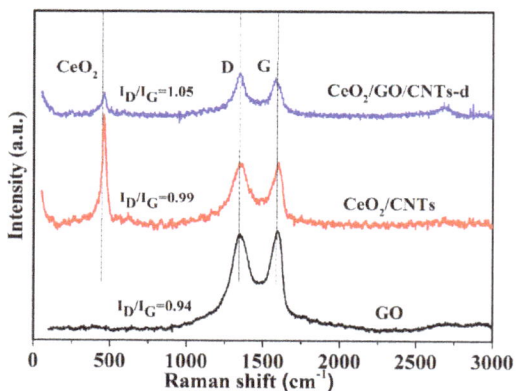

Figure 3. Raman spectra of GO, CeO_2/CNTs and CeO_2/GO/CNTs-d nanocomposites.

The elemental composition and chemical state of nanocomposites can be analyzed using XPS. The C1s XPS spectrum of CeO_2/CNTs nanocomposites shown in Figure 4a was deconvoluted into four peaks located at 289.2 eV (O=C–O), 286.5 eV (C=O), 285.6 eV (C–OH) and 284.8 eV (C–C), which were similar to the C1s XPS spectrum of CeO_2/GO/CNTs nanocomposites in Figure 4b [39]. From Figure 4c, the presence of the Ce–O bond can demonstrate the successful preparation of CeO_2/GO/CNTs nanocomposites via hydrothermal reaction. In Figure 4d, six Ce 3d peaks located at 917.4, 908.4, 901.5, 899.1, 886.3, and 882.9 eV, respectively, were assigned to $3d_{3/2}$ and $3d_{5/2}$ for Ce^{4+} final states. Moreover, two weak peaks located at 902.25 and 889.7 eV should be assigned to $3d_{3/2}$ and $3d_{5/2}$ for Ce^{3+} final states [33]. In combination with Figure 4b–d, it can be speculated that surfaces of both GO and CNTs were rich in oxygen-containing groups so that CeO_2 can be homogeneously compounded with them using a hydrothermal method.

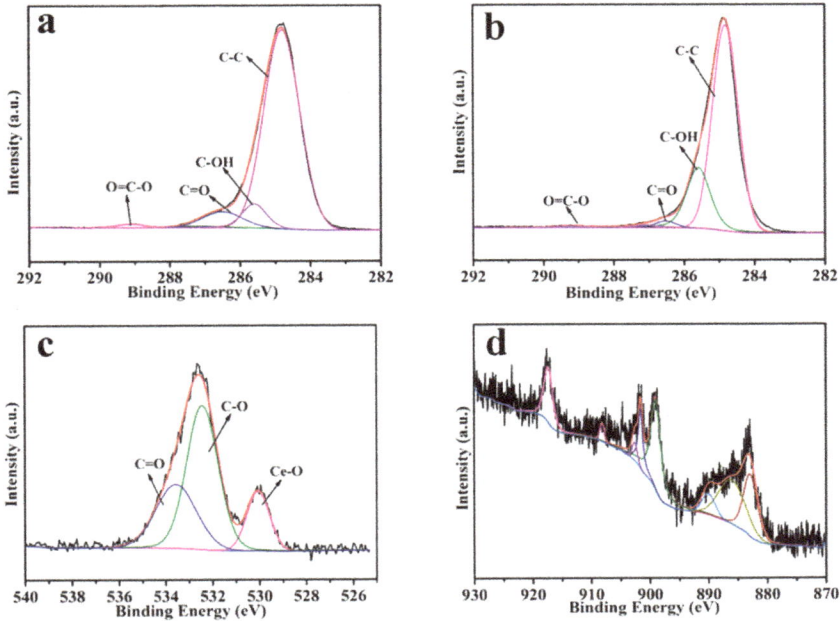

Figure 4. XPS survey spectrum for (**a**) C1s of CeO$_2$/CNTs nanocomposites; (**b**) C1s; (**c**) O1s; and (**d**) Ce3d spectrum for CeO$_2$/GO/CNTs-d nanocomposites.

The dispersion stability of lubricating additive is a significant factor that affects the tribological performance of lubricating oil. Paraffin oil used as the lubricant base oil in this study has a similar structure with petroleum and is more environmentally friendly than petroleum [40]. Figure 5 showed that 1 wt% CeO$_2$/GO/CNTs-d nanocomposites were dispersed in paraffin oil under ultrasonic agitation and placed in a water bath at 30 °C for observation. With the passage of time, the dispersion stability of nanocomposites in base oil was observed using a digital camera. We could clearly see that CeO$_2$/GO/CNTs-d nanocomposites possessed excellent dispersion and remained stable in the base oil on the twelfth day. The long-term dispersion stability of the nanocomposites in the base oil resulted from the uniform dispersion of CeO$_2$ nanoparticles on the GO/CNTs hybrids and the π-π bond interaction and rich oxygen-containing groups on the surface of nanocomposites, which could prevent the agglomeration between nanocomposites [30,31,41].

Generally, the friction performances of various nanomaterials used as additives and paraffin oil were characterized by the friction coefficient and wear scar diameters (WSD). Demonstrated in Figure 6 are variation of friction coefficient, average friction coefficient, and WSD of paraffin oil, 1 wt% of GO nanosheets, CeO$_2$ nanoparticles, CeO$_2$/CNTs nanocomposites, and CeO$_2$/GO/CNTs-d nanocomposites. Figure 6a shows that the paraffin oils filled with additives have superior stability compared with pure paraffin oil whose friction coefficient was increased with the passage of time. Additionally, the friction coefficient of paraffin oils filled with additives was lower than that of pure paraffin oil. It indicated that the addition of these additives can enhance the tribological performance of pure paraffin oil. Evidently, paraffin oils filled with CeO$_2$/GO/CNTs-d nanocomposites manifested the lowest friction coefficient. These results are consistent with average friction coefficient of abovementioned pure paraffin oil and paraffin oils filled with different additives shown in Figure 6b. WSD is of great significance for demonstrating the friction performance of paraffin oils. The trend of WSD of nanocomposites in Figure 6b is the same as the friction coefficient trend wherein the friction coefficient and WSD of paraffin oil filled with CeO$_2$/GO/CNTs-d nanocomposites decreased by 31.6%

and 37% compared with pure paraffin oil, respectively. The friction coefficient and WSD of paraffin oil filled with CeO_2/CNTs nanocomposites did not follow the improvement trend. The aggregation of CeO_2 nanoparticles on CNTs was observed in Figure 2 and the CeO_2/CNTs additives were prone to precipitate because of its poor compatibility with the lubricant base oil, which led to slightly influencing the improvement of tribological properties. Therefore, all these additives were effective for enhancing the tribological property of paraffin oil. In particular, the paraffin oil filled with CeO_2/GO/CNTs-d nanocomposites displayed excellent tribological properties. This can be explained by the fact that the two-dimensional lamellar structure of GO, the one-dimensional tubular CNTs, and the zero-dimensional spherical CeO_2 were uniformly compounded and achieved the synergistic lubrication effect [42] that the improvement of tribological properties of CeO_2/GO/CNTs-d nanocomposites was far superior to the single-component nanomaterials.

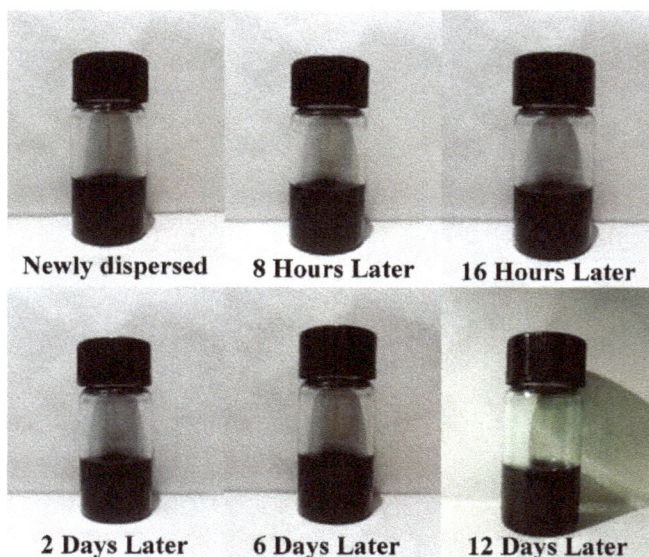

Figure 5. Digital photos of dispersion stability of 1 wt% CeO_2/GO/CNTs-d nanocomposites in paraffin oil at 30 °C.

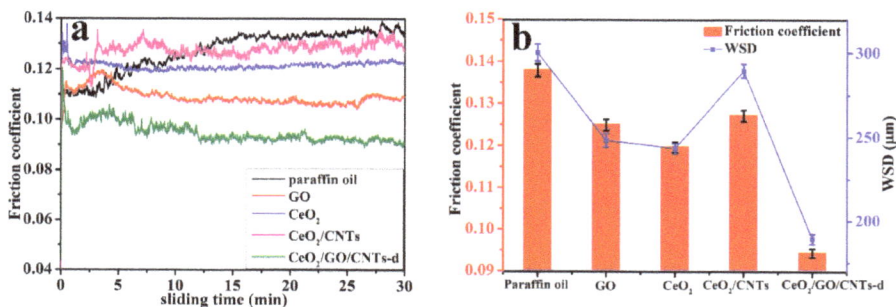

Figure 6. (**a**) Typical friction coefficient curves using the base oil and the oil with 1 wt% GO nanosheets, CeO_2 nanoparticles, CeO_2/CNTs nanocomposites, and CeO_2/GO/CNTs-d nanocomposites; and (**b**) comparisons of average friction coefficient and average WSD of steel disks using these oils over the entire duration of the test (load of 10 N, rotating speed of 300 rpm, test duration of 30 min).

Appl. Sci. **2019**, *9*, 170

Considering the effect of different weight ratios of CeO_2, GO, and CNTs in the nanocomposite on the friction properties of paraffin oil, the tribological behaviors of CeO_2/GO/CNTs nanocomposites with a content of 1 wt% in paraffin oil were investigated as lubricant additives in pure paraffin oil. In Figure 7a, the paraffin oil filled with CeO_2/GO/CNTs nanocomposites using different weight ratios of raw materials gave different tribological properties. Therein, the friction coefficients of paraffin oil filled with the CeO_2/GO/CNTs-a and CeO_2/GO/CNTs-e nanocomposites presented large fluctuations and tended to rise slowly. In contrast to the instability of friction coefficients of paraffin oil filled with CeO_2/GO/CNTs-a and CeO_2/GO/CNTs-e nanocomposites, paraffin oil filled with CeO_2/GO/CNTs-(b–d) nanocomposites not only expressed the better stability but also had the lower friction coefficients. From the curve of friction coefficient and WSD shown in Figure 7b, it can be observed that CeO_2/GO/CNTs-d nanocomposites had the lowest friction coefficient and the narrowest WSD compared with other additives. The comprehensive analysis signified that GO/CNTs hybrids decorated with uniform CeO_2 nanoparticles were achieved at a proper mass ratio of GO/CNTs and $Ce(NO_3)_3 \cdot 6H_2O$, which is an important point to reduce the wear rate of base oil. Therefore, CeO_2/GO/CNTs-d possessed the superior lubricating performance owing to the uniform distribution of CeO_2 nanoparticles on surface of hybrids and no obvious aggregation of among CeO_2, CNTs, and GO, which is consist with the TEM images shown in Figure 2.

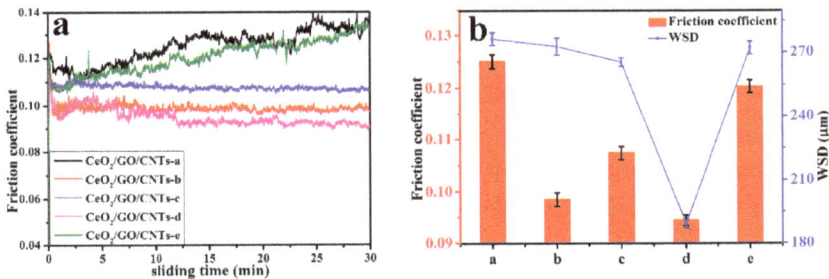

Figure 7. (**a**) Typical friction coefficient curves using the base oil with different nanocomposites; and (**b**) comparisons of average friction coefficient and average WSD of steel disks using these oils over the entire duration of the test (load of 10 N, rotating speed of 300 rpm, test duration of 30 min).

In order to obtain the appropriate content of CeO_2/GO/CNTs-d nanocomposites as a lubricant additive that improves the friction resistance of paraffin oil, the tribological performance of paraffin oil with different contents of CeO_2/GO/CNTs-d nanocomposites were investigated, as shown in Figure 8a,b. From the curve of friction coefficient with time in Figure 8a, it can be found that the CeO_2/GO/CNTs-d nanocomposites lubricant with a content of 0.5 wt% and 1 wt% were relatively stable. In particular, in Figure 7b, CeO_2/GO/CNTs-d nanocomposites with a content of 1 wt% manifested the lowest friction coefficient and the smallest WSD compared with other concentrations of additives. Initially, the friction coefficient and WSD declined with the increasing concentration of CeO_2/GO/CNTs-d. Base oil with low concentration (0.5 wt%) additives exhibited a higher friction coefficient and WSD due to the incomplete dispersion of additives in oil. While the concentration of CeO_2/GO/CNTs-d exceeded 1 wt%, the friction coefficient rose with the increasing concentrations of CeO_2/GO/CNTs-d, and the value of WSD was much more than 1 wt% CeO_2/GO/CNTs-d due to agglomeration. Therefore, 1 wt% CeO_2/GO/CNTs-d nanocomposites were the most suitable concentration, which makes the corresponding paraffin oil possess excellent friction resistance. The excellent lubricating performance was attributed to the perfect combination of the laminated GO nanosheets and the ball-bearing CeO_2 and CNTs [41–43].

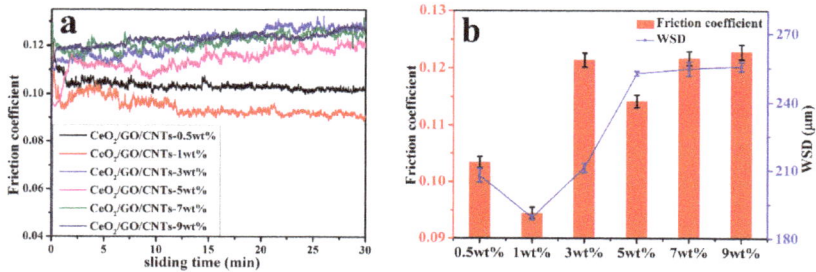

Figure 8. (a) Typical friction coefficient curves using the base oil with different contents of $CeO_2/GO/CNTs$-d nanocomposites; and (b) comparisons of average friction coefficient and average WSD of steel disks using these oils over the entire duration of the test (load of 10 N, rotating speed of 300 rpm, test duration of 30 min).

4. Conclusions

In this paper, 0-D CeO_2 nanoparticles, 1-D CNTs, and 2-D GO were used to formulate $CeO_2/GO/CNTs$ nanocomposites using a hydrothermal reaction and the CeO_2 nanoparticles were uniformly distributed on GO and CNTs. The effect of $CeO_2/GO/CNTs$ nanocomposites as a lubricating additive on the friction performance of paraffin oil were investigated. Compared with $CeO_2/GO/CNTs$-(a,b,c,e) nanocomposites containing the different weight ratios of CeO_2, GO, and CNTs, the paraffin oil with $CeO_2/GO/CNTs$-d nanocomposites was declared as having the lower friction coefficient and narrower WSD. Moreover, when the mass fraction was 1 wt%, as a lubricating oil additive, $CeO_2/GO/CNTs$-d manifested an excellent anti-friction performance. These results demonstrated that $CeO_2/GO/CNTs$ nanocomposite additives possessed valuable properties for improving the tribological properties of lubricating oils due to the synergistic effect among three different size dimensions.

Author Contributions: C.M., H.S. and L.G. conceived and designed the experiments; W.J., J.Q., and Y.J. performed the experiments; D.L. analyzed the data; Z.H. wrote the paper.

Funding: This research was funded by the National Natural Science Foundation of China [51875330, 51803081, 51103065], the Tribology Science Fund of State Key Laboratory of Tribology [SKLTKF17B08] and the Project National United Engineering Laboratory for Advanced Bearing Tribology [201806]. And the APC was funded Postgraduate Research & Practice Innovation Program of Jiangsu Province [KYCX18_2233].

Conflicts of Interest: The authors declare no conflict of interest.

References

1. Min, C.; Liu, D. Preparation of novel polyimide nanocomposites with high mechanical and tribological performance using covalent modified carbon nanotubes via Friedel-Crafts reaction. *Polymer* **2018**, *150*, 223–231. [CrossRef]

2. Wang, J.; Zhang, J.H. Study on Lubrication Performance of Journal Bearing with Multiple Texture Distributions. *Appl. Sci.* **2018**, *8*, 244. [CrossRef]

3. Wang, J.; Cai, M. Experimental Evaluation of Lubrication Characteristics of a New Type Oil-Film Bearing Oil Using Multi-Sensor System. *Appl. Sci.* **2017**, *7*, 28. [CrossRef]

4. Lee, K.; Hwang, Y.; Cheong, S. Understanding the role of nanoparticles in nano-oil lubrication. *Tribol. Lett.* **2009**, *35*, 127–131. [CrossRef]

5. Shahnazar, S.; Bagheri, S.; Hamid, S.B.A. Enhancing lubricant properties by nanoparticle additives. *Int. J. Hydrogen Energy* **2016**, *41*, 3153–3170. [CrossRef]

6. Bakunin, V.N.; Suslov, A.Y.; Kuzmina, G.N. Synthesis and application of inorganic nanoparticles as lubricant components—A review. *J. Nanopart. Res.* **2004**, *6*, 273–284. [CrossRef]

7. Li, B.; Wang, X.; Liu, W. Tribochemistry and antiwear mechanism of organic-inorganic nanoparticles as lubricant additives. *Tribol. Lett.* **2006**, *22*, 79–84. [CrossRef]

8. Min, C.; Liu, D.; Shen, C. Remarkable improvement of the wear resistance of poly(vinylidene difluoride) by incorporating polyimide powder and carbon nanofibers. *Appl. Phys. A Mater. Sci. Process.* **2017**, *123*, 638. [CrossRef]

9. Min, C.; Liu, D.; Shen, C. Unique synergistic effects of graphene oxide and carbon nanotube hybrids on the tribological properties of polyimide nanocomposites. *Tribol. Int.* **2018**, *117*, 217–224. [CrossRef]

10. Xue, M.; Wang, Z.; Yuan, F. Preparation of TiO_2/Ti_3C_2Tx hybrid nanocomposites and their tribological properties as base oil lubricant additives. *RSC Adv.* **2017**, *7*, 4312–4319. [CrossRef]

11. Xu, D.; Tong, Y.; Yan, T.T.; Shi, L.Y.; Zhang, D.S. N,P-Codoped Meso-/Microporous Carbon Derived from Biomass Materials via a Dual-Activation Strategy as High-Performance Electrodes for Deionization Capacitors. *ACS Sustain. Chem. Eng.* **2017**, *5*, 5810–5819. [CrossRef]

12. Li, W.; Cheng, Z.L.; Liu, Z. Novel Preparation of Calcium Borate/Graphene Oxide Nanocomposites and Their Tribological Properties in Oil. *J. Mater. Eng. Perform.* **2017**, *26*, 285–291. [CrossRef]

13. Wang, Z.; Ren, R.; Song, H. Improved tribological properties of the synthesized copper/carbon nanotube nanocomposites for rapeseed oil-based additives. *Appl. Surf. Sci.* **2018**, *428*, 630–639. [CrossRef]

14. Zhuoming, G.U.; Caixiang, G.U.; Shaoqing, F. Study on Lubricating Oils Containing Mixtures Additives of Nano-CaCO and Nano-Cu. *J. Mater. Eng.* **2007**, *21*, 19–22. [CrossRef]

15. Cadek, M.; Coleman, J.N.; Ryan, K.P. Reinforcement of polymers with carbon nanotubes: The role of nanotube surface area. *Nano Lett.* **2004**, *4*, 353–356. [CrossRef]

16. George, R.; Kashyap, K.T.; Rahul, R. Strengthening in carbon nanotube/aluminium (CNT/Al) composites. *Scr. Mater.* **2005**, *53*, 1159–1163. [CrossRef]

17. Curtin, W.A.; Sheldon, B.W. CNT-reinforced ceramics and metals. *Mater. Today* **2004**, *7*, 44–49. [CrossRef]

18. Gofman, I.; Zhang, B.; Zang, W. Specific features of creep and tribological behavior of polyimide-carbon nanotubes nanocomposite films: Effect of the nanotubes functionalization. *J. Polym. Res.* **2013**, *20*, 258. [CrossRef]

19. Chen, C.S.; Chen, X.H.; Xu, L.S. Modification of multi-walled carbon nanotubes with fatty acid and their tribological properties as lubricant additive. *Carbon* **2005**, *43*, 1660–1666. [CrossRef]

20. Ji, L.; Rao, M.; Zheng, H. Graphene oxide as a sulfur immobilizer in high performance lithium/sulfur cells. *J. Am. Chem. Soc.* **2017**, *133*, 18522–18525. [CrossRef]

21. Chen, D.; Feng, H.; Li, J. Graphene Oxide: Preparation, Functionalization, and Electrochemical Applications. *Chem. Rev.* **2012**, *112*, 6027–6053. [CrossRef] [PubMed]

22. Paredes, J.I.; Villar-Rodil, S.; Martinez-Alonso, A. Graphene Oxide Dispersions in Organic Solvents. *Langmuir* **2008**, *24*, 10560–10564. [CrossRef] [PubMed]

23. Kim, H.J.; Penkov, O.V.; Kim, D.E. Tribological Properties of Graphene Oxide Nanosheet Coating Fabricated by Using Electrodynamic Spraying Process. *Tribol. Lett.* **2015**, *57*, 27. [CrossRef]

24. Yang, J.; Xia, Y.; Song, H.; Chen, B.; Zhang, Z. Synthesis of the liquid-like graphene with excellent tribological properties. *Tribol. Int.* **2017**, *105*, 118–124. [CrossRef]

25. Zhang, S.; Yang, J.; Chen, B.; Guo, S.; Li, J.; Li, C. One-step hydrothermal synthesis of reduced graphene oxide/zinc sulfide hybrids for enhanced tribological properties of epoxy coatings. *Surf. Coat. Technol.* **2017**, *326*, 87–95. [CrossRef]

26. Chen, B.; Li, X.; Jia, Y.; Xu, L.; Liang, H.Y. Fabrication of ternary hybrid of carbon nanotubes/graphene oxide/MoS2 and its enhancement on the tribological properties of epoxy composite coatings. *Compos. Part A* **2018**, *115*, 157–165. [CrossRef]

27. Xin, Y.; Li, T.; Gong, D. Preparation and tribological properties of graphene oxide/nano-MoS_2 hybrid as multidimensional assembly used in the polyimide nanocomposites. *RSC Adv.* **2017**, *7*, 6323–6335. [CrossRef]

28. Zhang, X.; Luster, B.; Church, A.; Muratore, C.; Voevodin, A.A.; Kohli, P. Carbon nanotube-MoS2 composites as solid lubricants. *ACS Appl. Mater. Interfaces* **2009**, *1*, 735–739. [CrossRef]

29. Bai, G.Y.; Wang, J.Q.; Yang, Z.G.; Wang, H.G. Preparation of a highly effective lubricating oil additive—Ceria/graphene composite. *RSC Adv.* **2014**, *4*, 47096–47105. [CrossRef]

30. Cote, L.J.; Kim, J.; Tung, V.C.; Luo, J. Graphene oxide as surfactant sheets. *Pure Appl. Chem.* **2011**, *83*, 95–110. [CrossRef]

31. Kim, J.; Cote, L.J.; Kim, F.; Yuan, W. Graphene Oxide Sheets at Interfaces. *J. Am. Chem. Soc.* **2010**, *132*, 8180–8186. [CrossRef]

32. Marcano, D.C.; Kosynkin, D.V.; Berlin, J.M. Improved Synthesis of Graphene Oxide. *ACS Nano* **2010**, *4*, 4806–4814. [CrossRef]

33. Zhou, Y.Z.; Yang, J.; Cheng, X.N. Electrostatic self-Assembly of graphene-silver multilayer films and their transmittance and electronic conductivity. *Carbon* **2012**, *50*, 4343–4350. [CrossRef]

34. Yang, S.Y.; Lin, W.N.; Huang, Y.L. Synergetic effects of graphene platelets and carbon nanotubes on the mechanical and thermal properties of epoxy composites. *Carbon* **2011**, *49*, 793–803. [CrossRef]

35. Xu, H.F.; Li, H. The effect of CO-doped on the room-temperature ferromagnetism of CeO_2 nanorods. *J. Magn. Magn. Mater.* **2015**, *377*, 272–275. [CrossRef]

36. Ammar, M.R.; Galy, N.; Rouzaud, J.N. Characterizing various types of defects in nuclear graphite using Raman scattering: Heat treatment, ion irradiation and polishing. *Carbon* **2015**, *95*, 364–373. [CrossRef]

37. Pham, D.T.; Lee, T.H.; Luong, D.H. Carbon nanotube-bridged graphene 3D building blocks for ultrafast compact supercapacitors. *ACS Nano* **2015**, *9*, 2018–2027. [CrossRef]

38. Yamamoto, G.; Hashida, T.; Adachi, K.; Takagi, T. Tribological Properties of Single-Walled Carbon Nanotube Solids. *J. Nanosci. Nanotechnol.* **2008**, *8*, 665–2670. [CrossRef]

39. Zhang, J.; Chen, C.; Yan, W. Ni nanoparticles supported on CNTs with excellent activity produced by atomic layer deposition for hydrogen generation from hydrolysis of ammonia borane. *Catal. Sci. Technol.* **2016**, *6*, 2112–2119. [CrossRef]

40. Yousef, S.; Visco, A.; Galtieri, G. Wear behaviour of UHMWPE reinforced by carbon nanofiller and paraffin oil for joint replacement. *Mater. Sci. Eng. C* **2017**, *73*, 234–244. [CrossRef]

41. Song, H.; Wang, Z.; Yang, J. Tribological properties of graphene oxide and carbon spheres as lubricating additives. *Appl. Phys. A* **2016**, *122*, 933. [CrossRef]

42. Bai, G.Y.; Wang, J.Q.; Yang, Z.G. Self-assembly of ceria/graphene oxide composite films with ultra-long antiwear lifetime under a high applied load. *Carbon* **2015**, *84*, 197–206. [CrossRef]

43. Song, H.J.; Wang, Z.Q.; Yang, J.; Jia, X.H. Facile synthesis of copper/polydopamine functionalized graphene oxide nanocomposites with enhanced tribological performance. *Chem. Eng. J.* **2017**, *324*, 51–62. [CrossRef]

applied
sciences

MDPI

Article

The Rotating Flow of Magneto Hydrodynamic Carbon Nanotubes over a Stretching Sheet with the Impact of Non-Linear Thermal Radiation and Heat Generation/Absorption

Sher Muhammad [1,2], Gohar Ali [1], Zahir Shah [3,4,*], Saeed Islam [3] and Syed Asif Hussain [1,2]

[1] Department of Mathematics, Islamia College, Peshawar 25000, KP, Pakistan;
shermphil123@gmail.com (S.M.); gohar.ali@icp.edu.pk (G.A.); asif@cecos.edu.pk (S.A.H.)
[2] Department of Mathematics, Cecos University of Information Technology & Emerging Science, Peshawar 25000, KP, Pakistan
[3] Department of Mathematics, Abdul Wali Khan University, Mardan 23200, KP, Pakistan;
saeedislam@awkum.edu.pk
[4] Gandhara Institute of Science & Technology, South Canal Road, Peshawar 25000, KP, Pakistan
* Correspondence: zahir1987@gist.edu.pk; Tel.: +92-333-9198-823

Received: 13 February 2018; Accepted: 19 March 2018; Published: 22 March 2018

Abstract: The aim of this research work is to investigate the innovative concept of magnetohydrodynamic (MHD) three-dimensional rotational flow of nanoparticles (single-walled carbon nanotubes and multi-walled carbon nanotubes). This flow occurs in the presence of non-linear thermal radiation along with heat generation or absorption based on the Casson fluid model over a stretching sheet. Three common types of liquids (water, engine oil, and kerosene oil) are proposed as a base liquid for these carbon nanotubes (CNTs). The formulation of the problem is based upon the basic equation of the Casson fluid model to describe the non-Newtonian behavior. By implementing the suitable non-dimensional conditions, the model system of equations is altered to provide an appropriate non-dimensional nature. The extremely productive Homotopy Asymptotic Method (HAM) is developed to solve the model equations for velocity and temperature distributions, and a graphical presentation is provided. The influences of conspicuous physical variables on the velocity and temperature distributions are described and discussed using graphs. Moreover, skin fraction coefficient and heat transfer rate (Nusselt number) are tabulated for several values of relevant variables. For ease of comprehension, physical representations of embedded parameters such as radiation parameter (Rd), magnetic parameter (M), rotation parameter (K), Prandtl number (Pr), Biot number (λ), and heat generation or absorption parameter (Q_h) are plotted and deliberated graphically.

Keywords: Single-Walled Carbon Nanotube (SWCNT); Multi-Walled Carbon Nanotube (MWCNT); MHD; Casson model; stretching sheet; non-linear thermal radiation; HAM

1. Introduction

Single-walled carbon nanotubes (SWCNTs) and multi-walled carbon nanotubes (MWCNTs) are similar in certain aspects, but they also have some striking differences. SWCNTs are an allotrope of sp^2 hybridized carbon, and are similar to fullerenes. Single-walled carbon nanotubes (SWCNTs) have unique character due to their unusual structure. The structure of SWCNTs demonstrates significant optical and electronic features, tremendous strength and flexibility, and high thermal and chemical stability. As a result, SWCNTs are expected to dramatically impact several industries, particularly with respect to electronic displays, health care, and composites. MCWNTs are also suitable for many applications due to their high conductivity. Applications include electrostatic discharge protection in wafer processing and fabrication, plastic components for automobiles, plastics rendered conductive to enable the electrostatic spray painting of automobile body parts. The carbon nanotube (CNT) is composed of allotropes of carbon, and has a cylinder-shaped nanostructure. The molecules of carbon in cylindrical form have remarkable physiognomies that are useful for applied sciences, optics, physical sciences, fabric sciences, energy, health sciences, and manufacturing [1]. Terminology with respect to carbon nanotubes (CNTs) was first presented by Iijima [2] in 1991, who explored multi-walled carbon nanotubes (MWCNTs) using the Krastschmer and Huffman technique. Single-walled carbon nanotubes (SWCNTs) were reported by Donald Bethune in 1993 [3]. It is acknowledged that nanofluids will reduce the difficulties connected with materials with low thermal conductivities, like coal oil, water, engine oil, gasoline etc. Nanofluids contain a homogeneous combination of nano-size materials, with lengths ranging from 1 nm to 100 nm in a base liquid. Nanofluids have applications in automobile engine freezing, heat interchangers, solar heating water, bio-medicine, cooling of micro-electronic chips, and enhanced productivity of diesel engines. Expressions with respect to nanofluids were initially presented by Choi [4] using the inconsistent thermal conductivity of liquid (water) by depositing copper nanomaterial. Said et al. [5] reported that the SWCNT nano-liquids improve the thermal performance of solar collectors in place of water, in a context where one of the most manageable and contamination-free source of renewable energy is solar energy, and solar aerials. It can transform solar radiation into heat energy. In the textile industry, much attention has been paid to coating CNTs because of their uses in the sound industry, for example in megaphones and headsets. Xiao et al. [6] reported that very thin CNT films radiate lurid sound, and therefore prepared thin film CNT megaphones. Halelfadel et al. [7] examined the productivity of water-based nanofluid CNTs and deliberated the consequences of stumpy volume fraction nanoparticles in a range from 0.0055% to 0.278% on the physical properties (density, viscosity, thermal conductivity, and specific heat) of nanofluids. Aman et al. [8] investigated the influence of magnetohydrodynamics on Poiseuille flow and heat analysis of CNTs with a Casson fluid vertical channel. Further studies on carbon nanotubes can be appraised in [9–14].

Hayat et al. [15] investigated the three-dimensional (3D) rotating flow of carbon nanotubes with a Darcy–Frochheimer porous medium. Manevitch et al. [16] studied nonlinear optical vibrations of single-walled carbon nanotubes. Imtiaz et al. [17] discussed the convective flow of carbon nanotubes between rotating stretchable disks with thermal radiation effects in his research article. In an article by Tran et al. [18] the purification and dissolution of carbon nanotube fiber spun using the floating catalyst method were reported.

In applied sciences and engineering, flows with respect to nonlinear materials are essential. To explore the properties of stream and heat analysis, a number of rheological models are available for second-, third-, and fourth-grade fluid, micro polar fluid, Jaffery fluid, and Maxwell fluid models. One such type of general model is the Casson model [19] presented to calculate the flow behaviors and display yield stress. When the produced stress is greater than the shear stress, the Casson fluid acts similarly to a solid. Examples of these (Casson) fluids are chocolate, honey, blood, jelly, tomato sauce, and soup. Dash et al. [20] studied the performances of Casson liquids over a permeable medium restricted by a circular cylinder. Pramanik [21] inspected the impact of radiation on the Casson liquid stream and heat through a porous extending medium. Hassanan et al. [22] investigated the unsteady Casson fluid layer and Newtonian heat through an oscillating surface. Asma et al. [23] examined the magnetohydrodynamic (MHD) unsteady flow of Casson fluid over an oscillating plate fixed in porous media. Walicka and Falicki [24] studied the influence of Reynold number on the stream of Casson liquid. Nadeem et al. [25] deliberated three-dimensional flows with convective boundary conditions of a Casson fluid. Specific studies related to Casson liquid flow can be found in [26,27].

The flow behavior is profoundly affected by the strength and placement of applied magnetic field. The applied magnetic field affects the deferred nanoparticles and reforms their absorption inside the fluid, which substantially modifies the flow features of heat analysis. The analysis of magnetohydrodynamic (MHD) flow of an electrically directing liquid flow has several major uses in

Appl. Sci. **2018**, *8*, 482

engineering fields, such as in atomic reactor freezing, magnetohydrodynamic power producers, plasma studies, and the petroleum industry. Arial [28] studied the impact of MHDs on the two-dimensional (2D) flow of a gummy liquid imposing normal to the plane, and analyzed the influence of the Hartmann number. Ganapathirao and Ravindran [29] investigated the suction/injunction effect on 2D MHD steady flow with the chemical reaction along with mixed convection. Rahman et al. [30] and Srinivasacharya et al. [31] deliberated the impression of heat generation/absorption on 2D nanofluid flow and reported that the Nusselt number was enhanced with large values of the Biot number and the slip parameter. Hameed et al. [32] investigated the combined magnetohydrodynamic and electric field effect on an unsteady Maxwell nanofluid flow over a stretching surface under the influence of variable heat and thermal radiation. Recently, Shah et al. [33,34] studied the effects of hall current on three dimensional non-Newtonian nanofluids and micropolar nanofluids in a rotating frame.

In various engineering processes, the boundary film stream on a stretching/extending sheet is of great importance. In metallurgical and polymer productions, stretching is a significant phenomenon in the construction of polymer sheets, food processing, film coating, and drawing of copper wire. Crane [35] is the pioneering author who introduced the fluid flow by means of the stretching sheet. The work of Crane was expanded upon by many scholars, like Vajraelu and Roper [36] who investigated the stream of second-grade fluids and stretching sheets. Rosca and Pop [37] and Sajid et al. [38] deliberated the vicious unsteady motion due to a shrinking/stretching curved medium. Further research on the liquid stream due to the stretching sheet can be found in several studies [39,40]. Similarly the thermal radiation is significant in certain applications where radiant discharge depends on temperature and nanoparticle volume fraction. Pooya et al. [41] performed a thermal exploration of single phase nanofluid flow. Afify et al. [42] considered the impact of thermal radiation on 2D boundary layer flow. Brinkman [43] has investigated the viscosity of concentrated suspensions and solution. Xue [44] has discussed the model for thermal conductivity of CNTs based composites.

Here, the main aim is to deal the heat generation/absorption along with thermal radiation in a steady MHD 3D rotational stream of a Casson nanofluid with an entirely different base liquid. The movement of liquid is created by means of a stretching sheet. Two classes of carbon nanotubes (CNTs), specifically single-walled carbon nanotubes (SWCNTs) and multiple-walled carbon nanotubes (MWCNTs), are employed to explore the behavior of nanofluids. The acceptable transformation implies alterations the system of partial differential equations (PDEs) to a mandatory system of ordinary differential equations (ODEs).

The Homotopy Asymptotic Method (HAM) [45–48] involves calculation expansion for velocities and the energy profile. Here, the expressions of velocities and temperature are mathematically and graphically examined. Furthermore, the coefficient of skin friction and Nesselt number are explored numerically, and the results are discussed graphically.

2. Problem Formulation

We assume a three-dimensional, viscous, steady, MHD rotating Casson nanofluid flow comprising single- and multi-walled CNTs over a stretching sheet. The flows of SWCNTs and MWCNTs are taken under the influence of non-linear thermal radiation and heat absorption/generation. We adopt the coordinate structure in such a manner that the surface associated with the xy-plane and fluid is placed in $z > 0$. The stretching rate of the surface and the angular velocity of fluid are assumed as $c > 0$ and Ω, respectively. The surface temperature is a result of the convective heating process, which involves the hot fluid temperature T_f and coefficient of heat transfer h_f. The geometry of the problem shown in Figure 1.

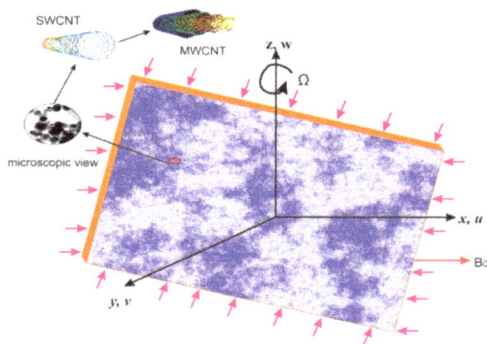

Figure 1. Schematic physical geometry.

The necessary constitutive model of a Casson fluid [19,22,24] is as follows:

$$\begin{aligned}
\frac{\tau_{mn}^{c.f}}{2e_{mn}} &= \mu_\gamma^{c.f} + \frac{p_y}{\sqrt{2\pi_d}}; \qquad \pi_d > \pi_{c.f}, \\
\frac{\tau_{mn}^{c.f}}{2e_{mn}} &= \mu_\gamma^{c.f} + \frac{p_y}{\sqrt{2\pi_{c.f}}}; \qquad \pi_d < \pi_{c.f}.
\end{aligned} \tag{1}$$

where $\tau_{mn}^{c.f}$ is the shear stress tensor of the Casson fluid, $\mu_\gamma^{c.f}$ is the Casson fluid plastic viscosity, π_d is the rate of deformation, and $\pi_{c.f}$ is the critical amount of current product that depends on the Casson fluid. We assume a situation in which a certain liquid requires the gradual growth of the the shear stress to sustain the constant rate of strain. In the case $\pi_d > \pi_{c.f}$ the Casson fluid possesses the kinematic viscosity $v^{c.f}$ as a function of $\mu_\gamma^{c.f}$, $\rho^{c.f}$, and Casson parameter β, such that

$$\begin{aligned}
\mu^{c.f} &= \mu_\gamma^{c.f} + \frac{p_y}{\sqrt{2\pi_d}}, \; p_y = \frac{\mu_\gamma^{c.f}\sqrt{2\pi_d}}{\beta}, \; v^{c.f} = \frac{\mu^{c.f}}{\rho^{c.f}}, \\
v^{c.f} &= \frac{\mu^{c.f}\left(1+\frac{1}{\beta}\right)}{\rho^{c.f}}, \; \frac{\partial \tau^{c.f}}{\partial z} = \left(1+\frac{1}{\beta}\right)u_{zz}.
\end{aligned} \tag{2}$$

Here $\mu^{c.f}$ represents the dynamic viscosity of the Casson fluid, $\beta = \frac{\mu_\gamma^{c.f}\sqrt{2\pi_d}}{p_y}$, and p_y shows the yield stress of the liquid. From the denotation of β describing the ratio of the product of $\mu_\gamma^{c.f}$ and $\sqrt{2\pi_d}$ to the yield stress p_y, it can be seen that the resistivity in liquid is greater if $\beta > 0$, and the liquid presents an inviscid behavior if $\beta < 0$, which eliminates the viscosity of the Casson fluid.

Taking into account the above declared rules, the primary equation takes the following form [8,25]:

$$u_x + v_y + w_z = 0, \tag{3}$$

$$\rho_{nf}\{u\,u_x + v\,u_y + w\,u_z - 2\Omega\,v\} = \mu_{nf}\left(1+\frac{1}{\beta}\right)(u_{zz}) - \sigma_{nf}\,B_0^2\,(u), \tag{4}$$

$$\rho_{nf}\{u\,v_x + v\,v_y + w\,v_z - 2\,\Omega\,u\} = \mu_{nf}\left(1+\frac{1}{\beta}\right)(v_{zz}) - \sigma_{nf}\,B_0^2\,(v), \tag{5}$$

$$u\,T_x + v\,T_y + w\,T_z = \alpha_{nf}\,(T_{zz}) - \frac{\left(q_z^{red}\right)}{(\rho c_p)_f} + \frac{Q_0(T-T_0)}{(\rho c_p)_f}. \tag{6}$$

The allied boundary conditions [25] are defined as

$$u_w = c\,x, \; v = 0, \; w = 0, \; -k_{nf}T_z = h_f\left(T_f - T\right) \text{ at } z = 0,$$
$$u = v = w = 0, \quad T = T_\infty, \quad \text{at} \quad z = \infty. \tag{7}$$

where u, v and w specify velocities in the x, y, z directions, respectively, ρ_{nf} represents the nanofluid density, μ_{nf} the nanofluid dynamic viscosity, and σ_{nf} the electrical conductivity of the nanofluid. T shows the local temperature, k_{nf} is the nanofluid thermal conductivity, and $(\rho c_p)_{nf}$ is the nanofluid specific heat capacity.

Brinkman [43] defined the dynamic viscosity of nano-liquid in term of base liquid as:

$$\mu_{nf} = \frac{\mu_f}{(1 - \varphi)^{2.5}}. \tag{8}$$

The density, specific heat capacity, and rate of thermal conductivities of the nanofluid [7,8] are expressed as:

$$\rho_{nf} = \rho_f(1 - \varphi) + \varphi\,\rho_{CNT}, \; \mu_{nf} = \frac{\mu_f}{(1-\varphi)^{2.5}},$$
$$\alpha_{nf} = \frac{\mu_f}{(\rho\,c_p)_{nf}}, \; (\rho\,C_p)_{nf} = (1 - \varphi)(\rho\,C_p)_f + \varphi(\rho\,C_p)_{CNT}, \tag{9}$$
$$\sigma_{nf} = \sigma_f\left(\frac{1 + 3(\sigma - 1)\,\varphi}{(\sigma + 2) - (\sigma - 1)\varphi}\right).$$

Here, the model by Xue [44] is implemented to calculate the thermal conductivities

$$\frac{k_{nf}}{k_f} = \frac{1 - \varphi + 2\varphi\left(\frac{k_{CNT}}{k_{CNT} - k_f}\,ln\left(\frac{k_{CNT} + k_f}{2\,k_f}\right)\right)}{1 - \varphi + 2\varphi\left(\frac{k_f}{k_{CNT} - k_f}\,ln\left(\frac{k_{CNT} + k_f}{2\,k_f}\right)\right)}. \tag{10}$$

where ρ_{CNT} shows the density of carbon nanotube, $(\rho c_p)_{CNT}$ denotes the specific heat of the carbon nanotube, and ρ_f, $\left(\rho c_f\right)_f$ represent the density and specific heat of base fluid, respectively. $q^{rad} = -\frac{4\sigma^*}{3k^*}\frac{\partial T^4}{\partial z}$ has been achieved using the Roseland approximation [43], where σ^* is the Stefan Boltzmann constant, and k^* is the absorption coefficient. We can observe that temperature difference in liquid motion is suitably negligible, so T^4 can be expressed in linear form using the Taylor's series, neglecting the higher terms: $T^4 = 4T_\infty^3 - 3T_\infty^4$. Therefore, we have $q^{rad} = -\frac{16\sigma^*}{3k^*}T^3\frac{\partial T}{\partial z}$.

Using the following suitable similarities variable [13],

$$u = c\,x\,f'(\eta), \; v = c\,x\,g(\eta), \; w = -\sqrt{c\,v_f}\,f(\eta),$$
$$\Theta(\eta) = T_\infty + \left(T_f - T_\infty\right)\Theta(\eta), \quad \eta = \left[\frac{c}{v_f}\right]^{\frac{1}{2}}. \tag{11}$$

Equation (3) is verified identically and Equations (4)–(10) take the following system when we use the similarities variables:

$$\left(1 + \frac{1}{\beta}\right)f'''(\eta) - \zeta_1\zeta_2\left[f(\eta)f''(\eta) + (f'(\eta))^2 + 2\,K\,g(\eta) - \zeta_4 M f'(\eta)\right] = 0, \tag{12}$$

$$\left(1 + \frac{1}{\beta}\right)g''(\eta) - \zeta_1\zeta_2\left[f(\eta)\,g'(\eta) + f'(\eta)\,g(\eta) - 2K\,f'(\eta) - \zeta_4 Mg(\eta)\right] = 0, \tag{13}$$

$$\frac{d}{d\eta}\left(\zeta_5 + \frac{4}{3}Rd[1 + (\Theta(\eta) - 1)\Theta(\eta)]^3\right)\Theta'(\eta) + \zeta_3 Pr\left[f(\eta)\Theta'(\eta) + Q_h\,\Theta(\eta)\right], \tag{14}$$

$$f(\eta) = g(\eta) = 0, \; f'(\eta) = 1, \; \Theta'(\eta) = \frac{k_f}{k_{nf}}(1 - \Theta(\eta))\lambda, \text{ at } \eta = 0,$$
$$f'(\eta) = g(\eta) = \Theta(\eta) = 0, \text{ at } \eta = \infty. \tag{15}$$

Here K represent the rotation parameter, M represents the magnetic parameter, Pr represents the Prandtl number, Q_h denotes the heat generation or absorption parameter, Rd represents radiation parameter, and λ is the Biot number, and are expressed as follows:

$$K = \frac{\Omega}{c}, M = \frac{\sigma_f B_0{}^2 g}{c}, Pr = \frac{(\rho\, c_p)_f}{k_f}, Q_h = \frac{Q}{c\,(\rho c_p)_f}, Rd = \frac{4\,\sigma^*\, T_\infty^3}{3k^* k_f}, \lambda = \frac{h_f}{k_f}\sqrt{\frac{\upsilon_f}{c}}. \tag{16}$$

$$\zeta_1 = (1-\varphi)^{2.5}, \zeta_2 = \left\{ (1-\varphi) + \frac{\rho_{CNT}\varphi}{\rho_f} \right\}, \zeta_3 = \left[(1-\varphi) + \frac{(\rho c_p)_{CNT}}{(\rho c_p)_f} \right],$$

$$\zeta_4 = \sigma_f \left(\frac{1+3(\sigma-1)\,\varphi}{(\sigma+2)-(\sigma-1)\varphi} \right), \zeta_5 = \frac{k_{nf}}{k_f} = \frac{1-\varphi+\left(2\frac{k_{CNT}}{k_{CNT}-k_f}\, ln\, \frac{k_{CNT}+k_f}{2\,k_f} \right)\varphi}{1-\varphi+\left(2\frac{k_f}{k_{CNT}-k_f}\, ln\, \frac{k_{CNT}+k_f}{2\,k_f} \right)\varphi}. \tag{17}$$

3. Physical Quantities of Interest

The physical quantities of interest such as Skin friction C_f is defined as $C_f = \left(1 + \frac{1}{\beta}\right)\frac{\mu_{nf}}{\rho_f u_w^2}(u_y)_{y=0}$ and local Nusselt number Nu_x, is defined as $Nu_x = \frac{xk_{nf}}{k_f(T_w - T_\infty)}(T_y)_{y=0}$, the dimensionless forms of these main design quantities are as follows:

$$\widetilde{C}_{fx} = \frac{1}{(1-\varphi)^{2.5}}\left(1 + \frac{1}{\beta}\right)f''(0), \widetilde{C}_{fy} = \frac{1}{(1-\varphi)^{2.5}}\left(1 + \frac{1}{\beta}\right)g'(0), Nu = -\frac{k_{nf}}{k_f}\Theta'(0). \tag{18}$$

4. Solution Methodology

In this section, the system of coupled nonlinear differential Equations (12)–(14), along with specific boundary conditions (Equation (15)) are solved analytically by using the Homotopy Analysis Method (HAM). The idea of the HAM was first presented by Liao [44,45] using the concept of homotopy. For solution development, the HAM scheme has certain advantages, such as the fact that it is free from small or large emerging parameters. The HAM offers an easy way to confirm the convergence of solution, and it provides liberty in terms of the right selection of the base function and the auxiliary parameter.

In the HAM scheme, the assisting parameter h is used to regulate and control the convergence of the solutions. The initial estimates are:

$$f_0(\eta) = 1 - e^{-x}, g_0(\eta) = e^{-x}, \Theta_0(\eta) = \frac{k_f\gamma}{k_f\gamma k_{nf}}e^{-x}. \tag{19}$$

The linear operators L are as follows:

$$L_f(f) = f_{\eta\eta\eta}, \quad L_g(g) = g_{\eta\eta}, \quad L_\Theta(\Theta) = \Theta_{\eta\eta}. \tag{20}$$

For the above stated differential operators, constants are shown as:

$$\begin{aligned} L_f(C_1 + C_2\eta + C_3\eta^2) &= 0, \\ L_g(C_4 + C_5\eta) &= 0, \\ L_\Theta(C_6 + C_7\eta) &= 0. \end{aligned} \tag{21}$$

4.1. Zeroth Order Deformation of the Problem

$U \in [0, 1]$ is an embedded parameter with auxiliary parameters h_f, h_g and h_Θ. The problem deforms for the zero order as follows:

$$(1 - U)L_g(f(\eta, U) - f_0(\eta)) = Uh_f N_f\{f(\eta, U), g(\eta, U)\}, \tag{22}$$

$$(1-U)L_g(g(\eta,U) - g_0(\eta)) = U\, h_g N_g\{f(\eta,U), g(\eta,U)\}, \tag{23}$$

$$(1-U)\, L_\Theta(\Theta(\eta,U) - \Theta_0(\eta)) = Uh_\Theta N_\Theta\{f(\eta,U), g(\eta,U), \Theta(\eta,U)\}. \tag{24}$$

The specific boundary conditions are obtained as:

$$\begin{aligned} f(0,U) = f'(0,U) = g(0,U) = \Theta(0,U) = 0, \\ f(1,U) = f'(1,U) = g(1,U) = \Theta(1,U) = 0. \end{aligned} \tag{25}$$

The resultant nonlinear operators are:

$$\begin{aligned} N_f(f(\eta;U),\, g(\eta;\,U)) = \left(1 + \tfrac{1}{\beta}\right)f_{\eta\eta\eta}\,(\eta;U) + \zeta_1\zeta_2\big[\{f_{\eta\eta}(\eta;U)\}^2 + f(\eta;U)f_{\eta\eta}(\eta;U) \\ +2K\,g(\eta;U) - \zeta_4\,M\,f_\eta(\eta;\,U)\big], \end{aligned} \tag{26}$$

$$\begin{aligned} N_g(g(\eta;U),\, f(\eta;U)) = \left(1 + \tfrac{1}{\beta}\right)g_{\eta\eta}\,(\eta;U) - \zeta_1\zeta_2\big[f_\eta\,\{(\eta;U)\,g(\eta;U)\} + \\ \{f(\eta;U)\,g_\eta(\eta;U)\} - 2K\,f_\eta(\eta;U) - \zeta_4 Mg(\eta;\,U)\big], \end{aligned} \tag{27}$$

$$\begin{aligned} N_\Theta(\Theta(\eta;U), f(\eta;U)) = \tfrac{d}{d\eta}\left(\zeta_5 + \tfrac{4}{3}Rd[1 + (\Theta(\eta) - 1)\Theta(\eta)]^3\right)\Theta_\eta(\eta;U) + \\ Pr\,\zeta_3\big[\{(f(\eta;U))\,\Theta_\eta(\eta;U)\} + Q_h\,\Theta\,(\eta;U)\big]. \end{aligned} \tag{28}$$

Taylor's series expresses $f(\eta;U)$, $g(\eta;U)$ and $\Theta(\eta;U)$. In terms of U we have:

$$\begin{aligned} f\,(\eta,U) = f_0(\eta) + \sum_{i=1}^{\infty} f_i(\eta), \\ g\,(\eta,U) = g_0(\eta) + \sum_{i=1}^{\infty} g_i(\eta), \\ \Theta\,(\eta,U) = \Theta_0(\eta) + \sum_{i=1}^{\infty} \Theta_i(\eta). \end{aligned} \tag{29}$$

where

$$f_i(\eta) = \frac{1}{i!}\{f_\eta(\eta,U)\}\bigg|_{U=0},\; g_i(\eta) = \frac{1}{i!}\{g_\eta(\eta,U)\}\bigg|_{U=0}, \Theta_i(\eta) = \frac{1}{i!}\{\Theta_\eta(\eta,U)\}\bigg|_{U=0} \tag{30}$$

4.2. ith-Order Deformation Problem

Now we differentiate the zeroth component equations *i*th time to achieve the *i*th order deformation equations with respect to U, so we get:

$$\begin{aligned} L_f(f_i(\eta) - \xi_i f_{i-1}(\eta)) = h_f\left(R_i^f(\eta)\right), \\ L_g(g_i(\eta) - \xi_i g_{i-1}(\eta)) = h_g\left(R_i^g(\eta)\right), \\ L_\Theta(\Theta_i(\eta) - \xi_i\Theta_{i-1}(\eta)) = h_\Theta(R_i^\Theta(\eta)). \end{aligned} \tag{31}$$

The resultant boundary conditions are:

$$\begin{aligned} f_i = f'_i = g_i = \Theta_i = 0, \quad at\; \eta = 0, \\ f_i = f' = g_i = \Theta_i = 0 \quad at\; \eta = 1. \end{aligned} \tag{32}$$

$$\begin{aligned} R_i^f(\eta) = \frac{d^3 f_{i-1}}{d\eta^3} + \zeta_1\zeta_2\left[\left(\sum_{j=0}^{i-1}\frac{d^2 f_{i-1}}{d\eta^2}\right)^2 + \left(\sum_{j=0}^{i-1}(f_{i-1-j})\frac{d^2 f_j}{d\eta^2}\right)\right. \\ \left. +2K\,g_{i-1} - \zeta_4 M\left(\sum_{j=0}^{i-1}\frac{d f_{i-1}}{d\eta}\right)\right], \end{aligned} \tag{33}$$

$$R_i^f(\eta) = \frac{d^2 g_{i-1}}{d\eta^2} + \zeta_1 \zeta_2 \left[\left(\sum_{j=0}^{i-1} \frac{df_{i-1}}{d\eta} g_j \right) + \left(\sum_{j=0}^{i-1} f_{i-1-j} \frac{dg_j}{d\eta} \right) - \zeta_4 M g_{i-1} - 2K \left(\sum_{j=0}^{i-1} \frac{df_{i-1}}{d\eta} \right) \right], \quad (34)$$

$$R_i^f(\eta) = \frac{d}{d\eta} \left(\zeta + \frac{4}{3} Rd[1 + (\Theta(\eta) - 1)\Theta(\eta)]^3 \right) \frac{d\Theta_{i-1}}{d\eta} +$$
$$Pr\, \zeta_3 \left[\sum_{j=0}^{i-1} f_{i-1-j} \frac{d\Theta_j}{d\eta} + Q_h \sum_{j=0}^{i-1} \Theta_{i-1-j} \right]. \quad (35)$$

where

$$\xi_i = \begin{cases} 1, & \text{if } U > 1 \\ 0, & \text{if } U \leq 1. \end{cases} \quad (36)$$

5. HAM Scheme Convergence

When we compute the series solutions of the velocity and temperature functions to use the HAM, the assisting parameters h_f, h_g, and h_Θ appear. These assisting parameters are responsible for adjusting the convergence of these solutions. In the possible region of velocities and temperature profiles, $f''(0)$, $g'(0)$, and $\Theta'(0)$ for the 15th-order approximation are plotted, for different values of the embedding parameter. The $\hbar$-curves consecutively display the valid region.

6. Results and Discussion

The present research has been carried out in order to study magnetohydrodynamics in the three-dimensional rotational flow of nanoparticles (single- and multi-walled carbon nanotubes) in the existence of non-linear thermal radiation along with heat generation or absorption based on the Casson fluid model over a stretching sheet. In this subsection we examine the physical outcomes of dissimilar embedded parameters with respect to the velocity distributions $f'(\eta)$, $g(\eta)$ and temperature distribution $\Theta(\eta)$. The influence of specific model quantities, such as K, M, Pr, Q_h, Rd, and λ (the rotation parameter, magnetic parameter, Pandtl number, heat generation or absorption parameter, radiation parameter, and Biot number, respectively) on $f'(\eta), g(\eta)$, and $\Theta(\eta)$ for SWCNTs and MWCNTs as nanoparticles is highlighted in the present discussion. Figure 1 shows the schematic flow geometry of the problem. Figures 2 and 3 present the convergence and probable choice of h-curve for $f''(0)$, $\Theta'(0)$ and $g'(0)$. The effects of model values such as $K, M, Pr, Q_h, Rd, \lambda$ on $f'(\eta), g(\eta)$ and $\Theta(\eta)$ are illustrated in Figures 4–14 and consequence are considered. Figure 4 demonstrates the consequence of rotation parameter K on the velocity profile $f'(\eta)$ for SWCNTs and MWCNTs. The increasing values of K depreciate the value of $f'(\eta)$ and result in a thinner boundary coating for SWCNTs and MWCNTs. Physically, when the value of rotation parameter K enlarges, the rate of rotation becomes dominant as compared to the stretching rate. Hence, there is a larger effect of rotation parameter K delivering resistance to the fluid stream. For the maximum value of the rotation parameter K, the values of velocity-filled $f'(\eta)$ become small near the boundary, and the layer becomes thinner, so the x-coordinate of velocity component is a decreasing function of the rotation parameter K. The similar impact for the high value of magnetic parameter M is shown in Figure 5 for both CNTs. It is obvious from the sketch that an increment in the magnetic parameter M depreciates the value of $f'(\eta)$. Usually, high values of magnetic parameter M improve the contradictory force (friction force), known as the Lorentz force. The corresponding force has the propensity to condense the value of $f'(\eta)$. The influence of volume fraction of nanoparticle φ on the velocity profile $f'(\eta)$ is exhibited in Figure 6 for SWCNTs and MWCNTs. We observe that increment in $f'(\eta)$ increases the quantity of φ for SWCNTs and MWCNTs. In fact, adding the nanoparticle φ enhances the energy and cohesive force between the atoms of the liquid, causing it to become frail and halt the faster liquid. Figure 7 illustrates the behavior of the Casson fluid parameter β on $f'(\eta)$ for SWCNTs and MWCNTs. With increasing values of the Casson fluid parameter β decreases the velocity field $f'(\eta)$ in the boundary layer. It is noted that to increases the value of the Casson fluid parameter β implies a reduction in the yield stress of the Casson liquid, and hence there is a clear acceleration in the motion of the boundary layer stream neighboring the stretching surface of the sheet. Moreover, it has been found that Casson fluids are similar to Newtonian fluids for large values of $\beta \to \infty$. The impact of nanoparticle volume fraction φ on $g(\eta)$ has been illustrated in Figure 8 for SWCNTs and MWCNTs. It is observed that the large value of φ decreases the transverse velocity $g(\eta)$ for SWCNTs and MWCNTs. In Figure 9 the velocity distribution $g(\eta)$ is intended for various values of rotation K. It is demonstrated that K has an energetic character in accelerating the flow in the y-plane. The suitably bulky value of K corresponds to an oscillatory pattern close to the wall of the sheet for SWCNTs and MWCNTs. The core reason is that the sheet is extended only along the x-plane, and due to rotation the liquid flow is solely possible in the y-direction. It can be observed that a high value of K reduces the magnitude of $g(\eta)$. Figure 10 shows that the features of the magnetic parameter M on $g(\eta)$ are qualitatively equivalent to those of K. Generally, the large value of M improves the resistance to the motion of liquid, which depreciates the value of $g(\eta)$. In Figure 11 it can be seen that the increment in Pr generates a fall in temperature $\Theta(\eta)$ due to the weaker thermal diffusivity. Physically, thermal layer thickness and $\Theta(\eta)$ form the reducing function of Pr. For the high Pr, the fluid retains low thermal conductivities, which causes the thinner thermal boundary layer. Consequently, the heat rate is increased. The fluids which have a small Pr possess excellent thermal conductivity. Figure 12 presents the impact of radiation parameter Rd on $\Theta(\eta)$ for SWCNTs and MWCNTs as nanoparticles. It seems that the increase in Rd boosts $\Theta(\eta)$, which is associated to the thickness of the boundary film. Basically, high Rd delivers more heat to functional nanofluids, which indicates an increment in $\Theta(\eta)$ in the presence of SWCNTs and MWCNTs. Figure 13 characterizes the temperature field $\Theta(\eta)$ for heat generation parameter Q_h. The value of $\Theta(\eta)$ and the thermal thickness of boundary layer are improved for a large value of Q_h. In this manner, the rate of heat is enhanced as the Q_h value is increased. Consequently, we note that the thermal thickness of boundary film is a function of Q_h.

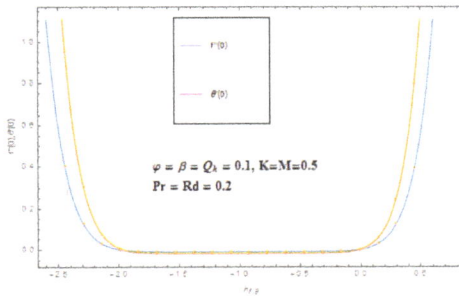

Figure 2. The combined h-curves of the velocity and heat profile with a 15-order approximation.

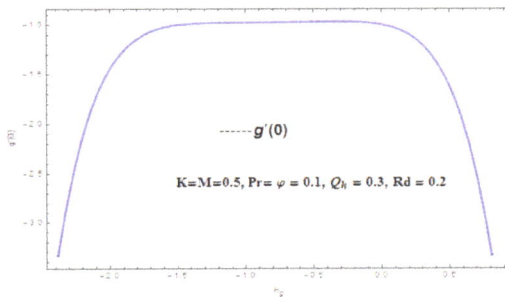

Figure 3. The h-curves of velocity $g(\eta)$ with a 15-order approximation.

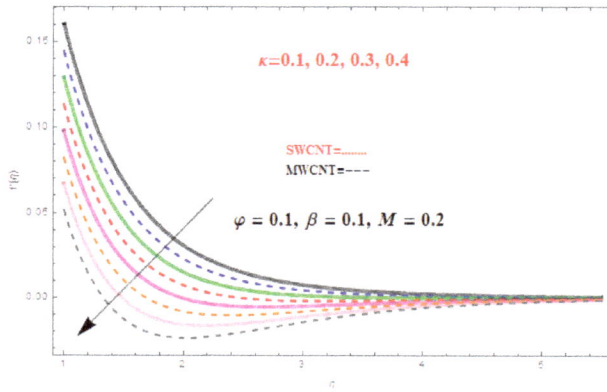

Figure 4. Effect of the rotation parameter K on $f'(\eta)$.

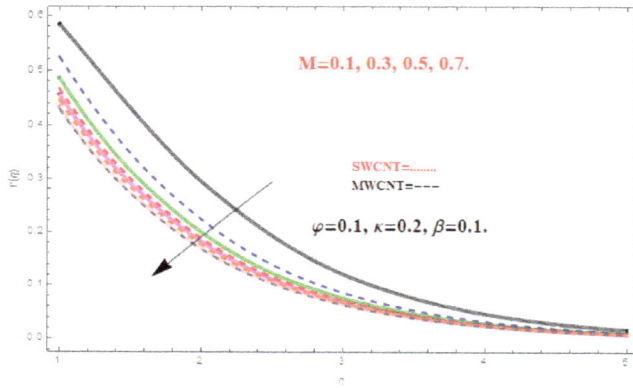

Figure 5. Effect of the magnetic parameter M on $f'(\eta)$.

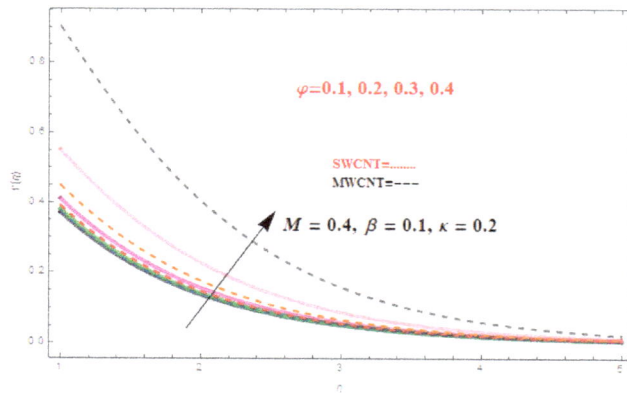

Figure 6. Effect of the volume fraction of the nanoparticle parameter φ on $f'(\eta)$.

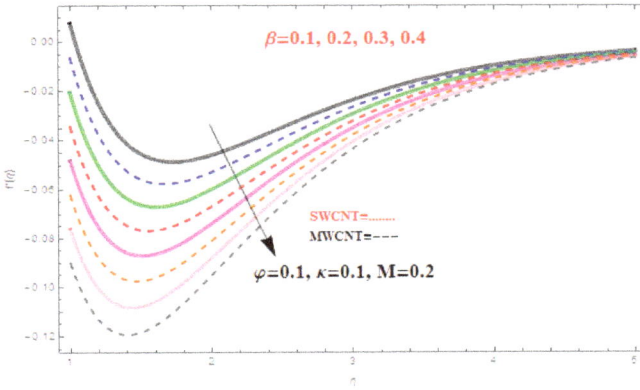

Figure 7. Effect of the Casson parameter β on $f'(\eta)$.

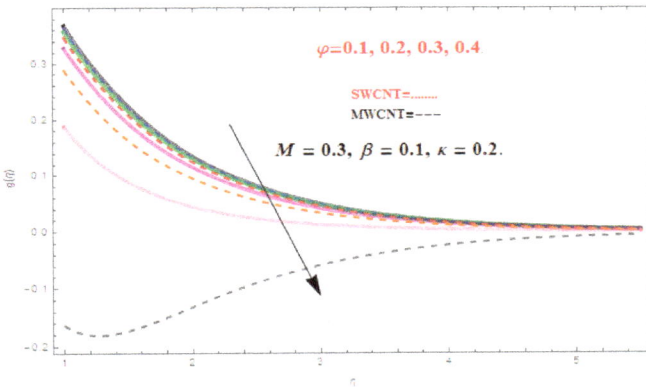

Figure 8. Effect of the volume fraction parameter φ on the velocity field $g(\eta)$.

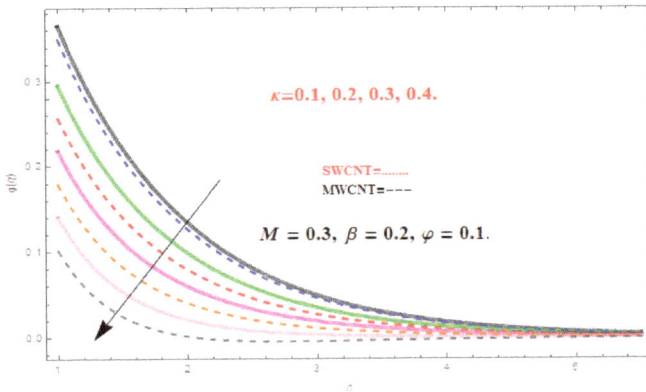

Figure 9. Effect of the rotation parameter K on the velocity field $g(\eta)$.

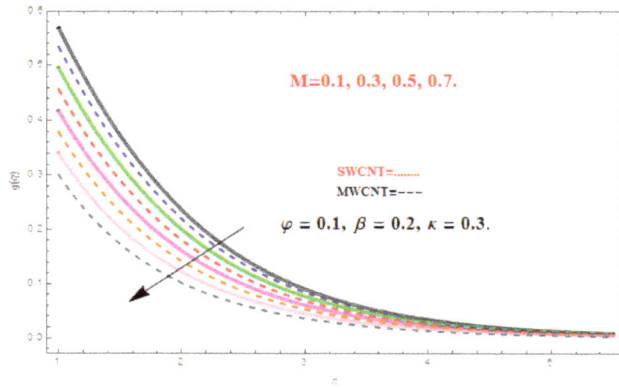

Figure 10. Effect of the magnetic parameter M on the velocity field $g(\eta)$.

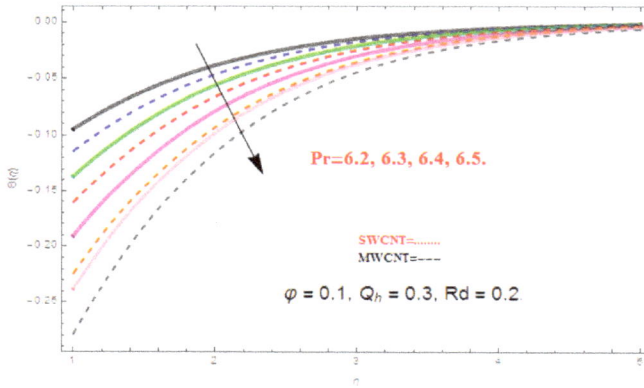

Figure 11. Effect of the Prandtl number Pr on the temperature field $\Theta(\eta)$.

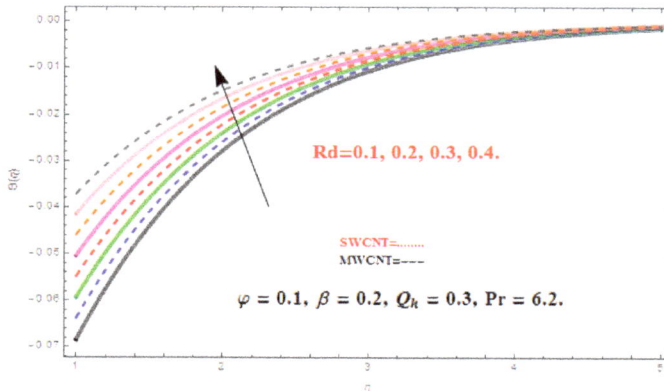

Figure 12. Effect of the radiation parameter Rd on the temperature field $\Theta(\eta)$.

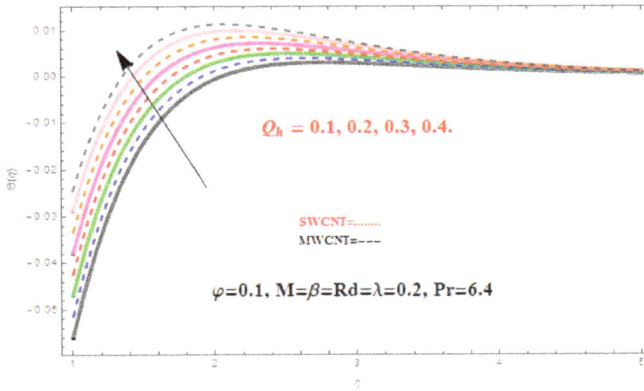

Figure 13. Effect of the heat generation/absorption parameter Q_h on the temperature field $\Theta(\eta)$.

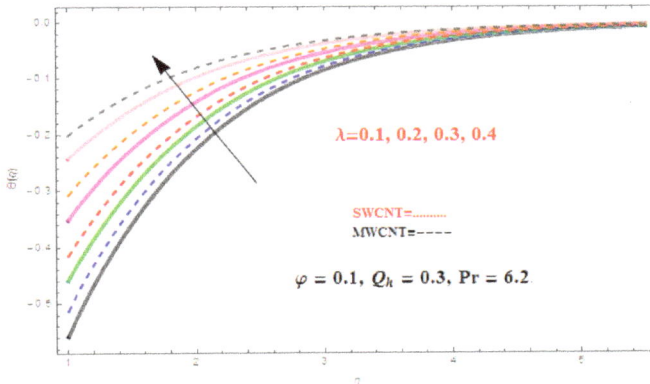

Figure 14. Effect of the Biot number λ on the temperature field $\Theta(\eta)$.

Variation in $\Theta(\eta)$ due to the Biot number λ is represented in the Figure 14 for both CNTs. The stronger value of λ leads to an enhanced temperature field $\Theta(\eta)$ and a thickness of the thermal film that allows the thermal impact to infiltrate deeper into the dormant liquid. In fact, λ depends on the heat transfer coefficient h_f, which is maximum for a massive value of λ, illustrating an increase in temperature.

Table Discussion

Table 1 illustrates various physical properties of certain base liquids (water, kerosene oil and engine oil) and nanoparticles such as SWCNTs and MWCNTs. Xue [44] calculated dissimilar values of effective thermal conductivities of CNT nanofluids for several values of the volume fraction ϕ of the CNTs shown in Table 2. It can be seen in Table 2 that the thermal conductivity is improved with inspiring volume fraction ϕ of CNTs. For the same values of ϕ, can be noted that the nanofluids with SWCNTs have higher effective thermal conductivities as compared to those with MWCNTs. The logic for this is the elevated value of thermal conductivity of SWCNTs (6600 Wm^{-1}/k^{-1}) as opposed to MWCNTs (3000 Wm^{-1}/k^{-1}), as given in Table 1.

Table 1. The thermo physical properties of carbon nanotubes (CNTs) of some base fluids. SWCNT: single-walled carbon nanotube; MWCNT: multi-walled carbon nanotube

Physical Properties		Density $\rho(\text{kg/m}^3)$	Thermal Conduct $k(\text{W/mk})$	Specific Heat $c_p(\text{kg}^{-1}/\text{k}^{-1})$
Base fluid	Water	997	0.613	4197
	Kerosene (lamp) oil	783	0.145	2090
	Engine oil	884	0.144	1910
Nanoparticles	SWCNT	2600	6600	425
	MWCNT	1600	3000	796

Table 2. Thermal conductivity values of CNTs with different volume fraction values.

Volume Fraction φ	Thermal Conductivity k_{nf} for SWCNT	Thermal Conductivity k_{nf} for MWCNT
0	0.145	0.145
0.01	0.174	0.172
0.02	0.204	0.2
0.03	0.235	0.228
0.04	0.266	0.257

Physical values such as the skin friction co-efficient $\left(C_f\right)$ and the Nusselt number (Nu) for the engineering purposes are calculated through Tables 3 and 4, respectively. Table 3 presents the numerical values of skin friction $\left(C_{fx}\right)$ and $\left(C_{fy}\right)$. It can be observed that the large values of M, φ and β increase the skin friction $\left(C_{fx}\right)$ and $\left(C_{fy}\right)$ of both SWCNTs and MWCNTs, where the large value of K reduces the skin friction $\left(C_{fx}\right)$ of both SWCNTs and MWCNTs and increases skin friction along the y-axis $\left(C_{fy}\right)$. Table 4 presents the numerical results of the Nusselt number (Nu). It can be noted that the large values of M and K increase the (Nu) of both SWCNTs and MWCNTs, where the large values of λ, Rd and Q_h reduce the Nusselt number of both SWCNTs and MWCNTs.

Table 3. Numerical data of the skin-friction coefficient for dissimilar values of M, K, φ, and β.

M	K	φ	fi	$-C_{fx}$		$-C_{fy}$	
				SWCNT	MWCNT	SWCNT	MWCNT
0.0	0.1	0.1	0.1	1.08611	1.03456	1.24004	1.14526
0.1				1.15356	1.12677	1.27582	1.19870
0.3				1.19726	1.14340	1.34626	1.24358
0.1	0.0			1.19564	1.04566	1.19564	1.13657
	0.3			0.98477	0.23561	1.44063	1.17452
	0.5			0.85341	0.02537	1.61105	1.32435
	0.1	0.0		0.96929	0.23409	1.10111	1.03452
		0.3		1.54113	1.45473	1.23251	1.10034
		0.5		1.68492	1.52435	1.69792	1.37342
		0.1	0.1	1.69211	1.57235	1.72340	1.22349
			0.3	1.73421	1.62345	1.88191	1.31902
			0.5	1.92345	1.81189	1.90823	1.59321

Appl. Sci. **2018**, *8*, 482

Table 4. Numerical data of Nusselt numbers for dissimilar values of M, K, λ, Rd, and Q_h when $Pr = 6.4$.

M	K	λ	Rd	Q_h	Nu_x SWCNT	Nu_x MWCNT
0.0	0.1	0.1	0.1	0.1	0.116964	0.231567
0.1					0.116974	0.232390
0.3					0.116988	0.233321
0.1	0.0				0.116965	0.134136
	0.3				0.116966	0.134342
	0.5				0.116967	0.134351
	0.1	0.3			0.116953	0.261532
		0.5			0.116943	0.261531
		0.8			0.116926	0.261530
		0.1	0.5		0.118451	0.156382
			1.0		0.120623	0.234521
			1.5		0.122502	0.267373
			0.1	0.5	0.116945	0.234536
				1.0	0.116928	0.198342
				1.5	0.116895	0.162435

7. Conclusions

We address the three-dimensional rotational flow of Casson fluids containing carbon nanotubes (SWCNTs and MWCNTs) as nanoparticles in the presence of thermal radiation along with heat generation over a stretching sheet. The effects of embedded parameters are observed and studied graphically. Moreover, the variation of the skin friction, Nusselt number, and their influences on the velocity and heat distributions are examined. The main points of the present study are shown below:

- For CNTs, large values of K (the rotation parameter) produce smaller velocities $f'(\eta)$ and $g(\eta)$, but the opposite tendency is detected for temperature profile $\Theta(\eta)$.
- Increasing the nanoparticle volume fraction φ into the working liquid boosted $f'(\eta)$, $g(\eta)$ and $\Theta(\eta)$ values for SWCNTs and MWCNTs.
- A strong magnetic parameter M reduces $f'(\eta)$ and $g(\eta)$, while increases the temperature $\Theta(\eta)$.
- The thermal layer rises through heat generation for SWCNTs and MWCNTs.
- The large values of Prandlt number Pr, reduces nanoparticle temperature $\Theta(\eta)$.
- Large values of Biot number λ boost temperature and thicken the boundary film concentration.
- The degree of coefficient of skin friction increases for larger φ values. The results show that skin friction in case of SWCNTs is higher than in MWCNTs.
- Heat transfer rate is controlled by large values of λ and Pr.
- Nusselt number is greater with SWCNTs as opposed to MWCNTs.

Acknowledgments: The authors are thankful to the anonymous reviewers and the Editor-in-Chief for the productive comments that led to positive enhancement in the paper.

Author Contributions: S.M. and Z.S. modelled and solved the problem. S.M. and G.A. wrote the manuscript. G.A. and S.I. thoroughly checked the mathematical modeling and English language corrections. S.I. and S.A.H. contributed in the results and discussions. All the corresponding authors finalized the manuscript after its internal evaluation.

Conflicts of Interest: The author declares that they have no competing interests.

References

1. Dai, L.; Chang, D.W.; Baek, J.B.; Lu, W. Carbon nanomaterials for advanced energy conversion and storage. *Small* **2012**, *8*, 1130–1166. [PubMed]
2. Iijima, S. Helical microtubules of graphitic carbon. *Nature* **1991**, *354*, 56–58. [CrossRef]
3. Ajayan, P.M.; Iijima, S. Capillarity-induced filling of carbon nanotubes. *Nature* **1993**, *361*, 333–334. [CrossRef]

4. Choi, S.U.S. *Enhancing Thermal Conductivity of Fluids with Nanoparticles, Developments and Applications of Non-Newtonian Flows*; Siginer, D.A., Wang, H.P., Eds.; American Society of Mechanical Engineers (ASME): New York, NY, USA, 1995; Volume 231, pp. 99–105.

5. Said, Z.; Saidur, R.; Rahim, N.A.; Alim, M.A. Analyses of energy efficiency and pumping power for a conventional flat plate solar collector using SWCNTs based nanofluid. *Energy Build.* **2014**, *78*, 1–9. [CrossRef]

6. Xiao, J.; Pan, X.; Guo, S.; Ren, P.; Bao, X. Toward fundamentals of confined catalysis in carbon nanotube. *J. Am. Chem. Soc.* **2015**, *137*, 477–480. [CrossRef] [PubMed]

7. Halelfadi, S.; Mare, T.; Estelle, P. Efficiency of carbon nanotubes water based nanofluids as coolants. *Exp. Therm. Fluid Sci.* **2014**, *53*, 104–110. [CrossRef]

8. Aman, S.; Khan, I.; Ismail, Z.; Salleh, M.Z.; Alshomrani, A.S.; Alghamdi, M.S. Magnetic field effect on Poiseuille flow and heat transfer of carbon nanotubes along a vertical channel filled with Casson fluid Citation. *AIP Adv.* **2017**, *7*, 015036. [CrossRef]

9. Hayat, T.; Hussain, Z.; Alsaedi, A.; Asghar, S. Carbon nanotubes effects in the stagnation point flow towards a nonlinear stretching sheet with variable thickness. *Adv. Powder Technol.* **2016**, *27*, 1677–1688. [CrossRef]

10. Wen, D.; Ding, Y. Effective thermal conductivity of aqueous suspensions of carbon nanotubes. *J. Thermophys. Heat Transf.* **2004**, *18*, 481–485. [CrossRef]

11. Mayer, J.; Mckrell, T.; Grote, K. The influence of multi-walled carbon nanotubes on single-phase heat transfer and pressure drop characteristics in the transitional flow regime of smooth tubes. *Int. J. Heat Mass Transf.* **2013**, *58*, 597–609. [CrossRef]

12. Kamli, R.; Binesh, A. Numerical investigation of heat transfer enhancement using carbon nanotube-based non-newtonian nanofluids. *Int. Commun. Heat Mass Transf.* **2010**, *37*, 1153–1157. [CrossRef]

13. Hussain, S.T.; Haq, R.U.; Khan, Z.H.; Nadeem, S. Water driven flow of carbon nanotubes in a rotating channel. *J. Mol. Liq.* **2016**, *214*, 136–144. [CrossRef]

14. Aman, S.; Khan, I.; Ismail, Z.; Salleh, M.Z.; Al-Mdallal, Q.M. Heat transfer enhancement in free convection flow of CNTs Maxwell nanofluids with four different types of molecular liquids. *Sci. Rep.* **2017**, *7*, 1–13. [CrossRef] [PubMed]

15. Hayat, T.; Haider, F.; Muhammad, T.; Alsaedi, A. Three Dimensional rotating flow of carbon nanotubes with Darcy-Frochheimer porous medium. *PLoS ONE* **2017**, *12*, e0179576. [CrossRef] [PubMed]

16. Manevitch, L.I.; Smirnov, V.V.; Strozzi, M.; Pellicano, F. Nonlinear optical vibrations of single-walled carbon nanotubes. *Int. J. Nonlinear Mech.* **2017**, *94*, 351–361. [CrossRef]

17. Imtiaz, M.; Hayat, T.; Alsaedi, A.; Ahmad, B. Convective flow of carbon nanotubes between rotating stretchable disks with thermal radiation effects. *Int. J. Heat Mass Transf.* **2016**, *101*, 948–957. [CrossRef]

18. Tran, T.Q.; Headrick, R.J.; Bengio, E.A.; Myint, S.M.; Khoshnevis, H.; Jamali, V.; Duong, H.M.; Pasquali, M. Purification and Dissolution of carbon nanotube fiber SPUM from floating catalyst method. *ACS Appl. Mater. Interfaces* **2017**, *9*, 37112–37119. [CrossRef] [PubMed]

19. Casson, N. *A Flow Equation for Pigment Oil Suspensions of the Printing Ink Type, Rheology of Disperse Systems*; Pergamon Press: New York, NY, USA, 1959; pp. 84–104.

20. Dash, R.K.; Mehta, K.N.; Jayaraman, G. Effect of yield stress on the flow of a Casson fluid in a homogeneous porous medium bounded by a circular tube. *Appl. Sci. Res.* **1996**, *57*, 133–149. [CrossRef]

21. Pramanik, S. Casson fluid flow and heat transfer past an exponentially porous stretching surface in presence of thermal radiation. *Ain Shams Eng. J.* **2014**, *5*, 205–212. [CrossRef]

22. Hussanan, A.; Salleh, M.Z.; Tahar, R.M.; Khan, I. Unsteady boundary layer flow and heat transfer of a Casson fluid past an oscillating vertical plate with Newtonian heating. *PLoS ONE* **2014**, *9*, e108763. [CrossRef] [PubMed]

23. Asma, K.; Khan, I.; Arshad, K.; Sharidan, S. Unsteady MHD free convection flow of Casson fluid past over an oscillatin vertical plate embedded in a porous medium. *Eng. Sci. Technol. Int. J.* **2015**, *18*, 309–317.

24. Walicka, A.; Falicki, J. Reynolds number effects in the flow of an electro rheological fluid of Casson type between fixed surface of revolution. *Appl. Math. Comput.* **2015**, *250*, 639–649.

25. Nadeem, S.; Haq, R.U.; Noreen, A.S. MHD three-dimensional boundary layer flow of Casson nanofluid past a linearly stretching sheet with convective boundary condition. *IEEE Trans. Nanotechnol.* **2014**, *13*, 109–115. [CrossRef]

26. Mukhopadhyay, S.; Bhattacharyya, K.; Hayat, T. Exact solutions for the flow of Casson fluid over a stretching surface with transpiration and heat transfer effects. *Chin. Phys. B* **2013**, *22*, 114701. [CrossRef]

27. Mukhopadhyay, S.; Mandal, I.S. Boundary layer flow and heat transfer of a Casson fluid past a symmetric porous wedge with surface heat flux. *Chin. Phys. B* **2014**, *23*, 044702. [CrossRef]
28. Ariel Hiemenz, P.D. Flow in hydro magnetics. *Acta Mech.* **1994**, *103*, 31–43.
29. Ganapathirao, M.; Ravindran, R. Non-uniform slot suction/injection into mixed convective MHD flow over a vertical wedge with chemical reaction. *Procedia Eng.* **2015**, *127*, 1102–1109. [CrossRef]
30. Rahman, M.M.; Al-Lawatia, M.A.; Eltayeb, I.A.; Al-Salti, N. Hydromagnetic slip flow of water based nanofluids past a wedge with convective surface in the presence of heat generation (or) absorption. *Int. J. Therm. Sci.* **2012**, *57*, 172–182. [CrossRef]
31. Srinivasacharya, D.; Mendu, U.; Venumadhav, K. MHD boundary layer flow of a nanofluid past a wedge. *Procedia Eng.* **2015**, *127*, 1064–1070. [CrossRef]
32. Hammed, H.; Haneef, M.; Shah, Z.; Islam, S.; Khan, W.; Muhammad, S. The Combined Magneto hydrodynamic and electric field effect on an unsteady Maxwell nanofluid Flow over a Stretching Surface under the Influence of Variable Heat and Thermal Radiation. *Appl. Sci.* **2018**, *8*, 160. [CrossRef]
33. Shah, Z.; Gul, T.; Khan, A.M.; Ali, I.; Islam, S. Effects of hall current on steady three dimensional non-newtonian nanofluid in a rotating frame with brownian motion and thermophoresis effects. *J. Eng. Technol.* **2017**, *6*, 280–296.
34. Shah, Z.; Islam, S.; Gul, T.; Bonyah, E.; Khan, M.A. The electrical MHD and hall current impact on micropolar nanofluid flow between rotating parallel plates. *Results Phys.* **2018**. [CrossRef]
35. Crane, L.J. Flow past a stretching plate. *Z. Angew. Math. Phys.* **1970**, *21*, 645–647. [CrossRef]
36. Vajravelu, K.; Roper, T. Flow and heat transfer in a second grade fluid over a stretching sheet. *Int. J. Nonlinear Mech.* **1999**, *34*, 1031–1036. [CrossRef]
37. Rosca, N.C.; Pop, I. Unsteady boundary layer OW over a permeable curved stretching/shrinking surface. *Eur. J. Mech. B* **2015**, *51*, 61–67. [CrossRef]
38. Sajid, M.; Ali, N.; Javed, T.; Abbas, Z. Stretching a curved surface in a viscous fluid. *Chin. Phys. Lett.* **2010**, *27*, 024703. [CrossRef]
39. Khan, W.A.; Pop, I. Boundary-Layer Flow of a Nanouid Past a Stretching Sheet. *Int. J. Heat Mass Transf.* **2010**, *53*, 2477–2483. [CrossRef]
40. Hassani, M.; Tabar, M.M.; Nemati, H.; Domairry, G.; Noori, F. An analytical solution for boundary layer flow of a Nano liquid past a stretching sheet. *Int. J. Therm. Sci.* **2011**, *50*, 2256–2263. [CrossRef]
41. Pooya, M.; Rad, C. The Effect of Thermal Radiation on Nanouid Cooled Microchannels. *J. Fusion Energy* **2009**, *28*, 91–100.
42. Afify, A.; Seddeek, M.A.; Bbazid, M.A.A. Radiation effects on Falkner-Skan flow of a nanouid past a wedge in the present of non-uniform heat source/sink. *Meccanica* **2011**, submitted.
43. Brinkman, H.C. The viscosity of concentrated suspensions and solution. *J. Chem. Phys.* **1952**, *20*, 571–581. [CrossRef]
44. Xue, Q. Model for thermal conductivity of carbon nanotube-based composites. *Phys. B Condens. Matter* **2005**, *368*, 302–307. [CrossRef]
45. Liao, S.J. On the homotopy analysis method for nonlinear problems. *Appl. Math. Comput.* **2007**, *147*, 499–513. [CrossRef]
46. Liao, S.J. Comparison between the homotopy analysis method and homotopy perturbation method. *Appl. Math. Comput.* **2005**, *169*, 1186–1194. [CrossRef]
47. Abbasbandy, S. The application of homotopy analysis method to solve a generalized Hirota-Satsuma Coupled KdV equation. *Phys. Lett. A* **2007**, *361*, 478–483. [CrossRef]
48. Zhen, W.; Li, Z.; Qing, Z.H. Solitary solution of discrete mKdV equation by homotopy analysis method. *Commun. Theor. Phys.* **2008**, *49*, 1373–1378. [CrossRef]

![applied sciences logo] *applied sciences*

MDPI

Article

Thermal Analysis of Nanofluid Flow over a Curved Stretching Surface Suspended by Carbon Nanotubes with Internal Heat Generation

Fitnat Saba [1], Naveed Ahmed [1], Saqib Hussain [2], Umar Khan [2,*], Syed Tauseef Mohyud-Din [3] and Maslina Darus [4]

[1] Department of Mathematics, Faculty of Sciences, HITEC University, Taxila Cantt 47080, Pakistan; fitnat_saba89@gmail.com (F.S.); nidojan@gmail.com (N.A.)
[2] Department of Mathematics, COMSATS Institute of Information Technology, Abbottabad 22060, Pakistan; saqib_math@yahoo.com
[3] Center for Research (CFR), University of Islamabad (UoI), Islamabad 44000, Pakistan; syedtauseefs@hotmail.com
[4] School of Mathematical Sciences, Faculty of Sciences and Technology, Universiti Kebangsaan Malaysia, Bangi 43600, Selangor, Malaysia; maslina@ukm.edu.my
* Correspondence: umarkhan@ciit.net.pk or umar_jadoon4@yahoo.com; Tel.: +92-332-890-2728

Received: 19 December 2017; Accepted: 12 February 2018; Published: 8 March 2018

Abstract: We have investigated a two-dimensional radiative flow of a boundary layer nature. The fluid under consideration is carbon nanotube (CNT)-based nanofluid and it flows over a curved surface. The heat transfer through the flow is analyzed under the influence of internal heat generation. Water (base fluid) along with single or multi-walled carbon nanotubes is taken to compose the nanofluid. After introducing the suitable similarity variables, the consequent equations are reduced to a system of nonlinear ordinary differential equations. The solution to the system is computed by using the shooting method accompanied by Runge–Kutta–Fehlberg algorithm. Various parameters, emerging in the governing equations, influences the flow and heat transfer distribution. These changes are captured and portrayed in the form of graphs. The changes in local rate of heat transfer and skin friction coefficient are also enlisted. To ensure the correctness of applied numerical scheme, the results are compared with some already existing studies.

Keywords: water based nanofluid; carbon-nanotubes; boundary layer; heat generation; thermal radiation; curved stretching sheet; numerical solution

1. Introduction

Nanomaterials have revolutionized many industrial and household appliances. Their use is gaining considerable popularity, and the fields of science associated with them enjoy a similar repute. Nanotechnology and its consequent products have many useful applications in various fields of science and engineering, from enhancing thermal properties of traditional fluids to construction of an effective drug delivery system. Other examples include electronics, life sciences, chemical synthesis, fuel cells, medical sciences, microsystems including mechanical and electrical components, etc. Nanofluids are formed by the addition of nano-sized particles in base fluids, which traditionally have poor thermal properties. These additives work as boosters and the resulting substance, i.e., nanofluid, bears some remarkable thermal properties [1–5]. Several mathematical models describe the properties of nanofluids. Choi [6] was the pioneer who witnessed and announced the existence of the tiny particles and their effectiveness. He examined through several experiments that the addition of nanometer sized particles of various nature exceptionally enhance the thermal properties of base fluids. Later, Choi et al. [7] presented a mathematical model which provides some new insights related to the field of nanotechnology. Following in his footsteps, many researchers carried out several research activities. Subsequently, several models have been proposed for different physical and industrial problems. One of these models is presented by Buongiorno [8]. His model considers both thermophoretic and Brownian motion effects. Hamilton and Crosser [9] formulated another model which describes the effects of shapes of nanoparticles. Some other famous models include Maxwell's [10] and Xue's [11] models. Khan and Pop utilized Buongiorno's concept to model a boundary layer flow over a stretching sheet [12]. There has been several efforts to use these models in different situations and geometries. Some of the most recent can be found in [13–20].

In 1991, Iijima [21] introduced a cylindrical member of fullerenes family named as carbon nanotubes (CNTs). They possess unique electronic, mechanical, chemical, optical, catalytic, adsorption, and transport properties, which make them very valuable for a wide range of applications [22,23]. They are mainly categorized in two modules, one Single-walled carbon nanotubes (SWCNTs) and the other multi-walled carbon nanotubes (MWCNTs). The term single refers to a single layer or wall that holds all the particles together, whereas the term multi-walled refers to a bunch of nested tubes whose diameter is continuously growing. It has been observed that the enhancement of thermal properties of base fluid using CNTs as nanoparticles is remarkable. Murshed et al. [24] showed that CNTs produce six times better thermal conductivity, at room temperature, as compared to other substances. The role of CNTs in enhancement of thermal properties of the base fluid has been highlighted by many scientists, in different scenarios and geometries [25–28].

During the last few years, researchers are showing a keen interest in understanding the flows of boundary layer nature over a dilating or a squeezing curved sheet. These types of flows bear commendable uses in different engineering and industrial processes such as production of papers, polymer sheet, manufacturing of rubber and plastic sheets, glass fibers, wire coating, food manufacturing, etc. In 1961, Sakiadis [29] was the first to come up with the idea of boundary layer flows. He presented the flow over a stretching surface. Crane [30] extended his idea for both linear and exponentially stretching sheets. He derived a closed form solution for the viscid flow caused by the stretching of a sheet. McLeod and Rajagopal [31] examined the uniqueness of the exact solution [30]. Since the pioneering work, flow problems related to a stretching surface under different situations have been analyzed by many scientists. Some of them can be found in References [32–37].

The energy crisis has become a major cause of concern in modern world. To meet a growing demand for energy, a new wave of innovative models is required that can replace the traditional ones. The resources for renewable energy, including solar and wind power, are now well placed to contribute to energy requirements in both mature and developing economies. Thermal radiation gains a great significance for such processes where the systems are running at very high temperatures, e.g., heating and cooling chambers, solar power technology, electrical power generation, thermal energy storage, nuclear power plants and many other industrial areas. Due to these important applications, many authors have contributed in studying the radiative flows of different fluids in various geometrical configurations [38,39].

Besides conduction and radiation, another means of heat transfer, i.e., convection, also plays a pivotal role in many real-life problems. Heat transfer via convection has been a core subject of many scientific investigations. Its role is very prominent in a wide range of energy-related applications such as automobile radiators, lubricants and coolants in various mechanical processes, etc. An asymptotic solution, by involving homotopy analysis method, to the stagnation point nanofluid flow (with a heat generation source) over a dilating permeable sheet was explained by Malvandi et al. [40]. Tsai et al. [41] focused on the influence of flow and heat transfer coupled with non-uniform heat source over an unsteady stretching surface. Pal [42] examined the radiative flow over an unsteady expanding surface with variable heat generating/consuming mechanism.

Previously, studies on the flows over flat surfaces were carried out using Cartesian coordinate systems. Recently, Sajid et al. [43] came up with an idea of curved surface and modeled the flow phenomena using a curvilinear coordinate system. Abbas with his coworkers [44] followed their approach and applied it to present a thermal analysis of the flow over a bent sheet. In 2016, Abbas et al. [45] studied the hydromagnetic and radiative slip flow of a nanofluid over a curved stretching surface along with heat generation. Most recently, a considerable interest has been shown by many authors in studying the flow over curved surfaces [46–48]. These recent efforts by renowned scholars have motivated us to carry out this work. We have examined the heat transfer in a radiative boundary layer flow, heat generation, of a carbon nanotube (CNT)-based nanofluid over a curved stretching sheet. Shooting method followed by Runge–Kutta–Fehlberg scheme is employed to obtain a numerical solution of the problem. Consequently, a comprehensive graphical description (highlighting the effects of various parameters on the velocity, pressure and temperature profiles) of the flow behavior is presented.

2. Formulation of the Problem

We consider a laminar flow of an incompressible nanofluid over a curved stretching sheet at $r^* = \mathcal{R}^*$ (see Figure 1). The flow is steady and two dimensional. The nanofluid under consideration is composed of CNT nanoparticles added to a base fluid (water). Since we are dealing with a curved sheet that is being stretched in opposite direction by two forces of same magnitude, we use curvilinear coordinates (r^*, s^*). The linear velocity by which the sheet is being stretched is taken to be $\acute{u} = \acute{a}s^*$, where $\acute{a}$ is a positive constant. The temperature of the sheet is represented by $\mathcal{T}^*_w$ ($\mathcal{T}^*_w(s^*) = \acute{A}(s^*/\ell)$; where $\acute{A}$ is a constant). The free stream temperature is denoted by $\mathcal{T}^*_\infty$ and it is assumed that $\mathcal{T}^*_\infty < \mathcal{T}^*_w$.

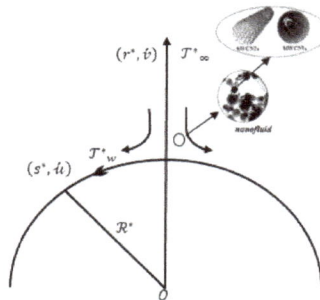

Figure 1. Physical illustration of the flow problem.

In context of the aforementioned coordinate scheme and the assumptions along with the boundary layer approximation, the equations representing the conservative balance for mass, momentum and energy are given as follows:

$$\frac{\partial}{\partial r^*}(Y\acute{v}) + \mathcal{R}^* \frac{\partial \acute{u}}{\partial s^*} = 0 \tag{1}$$

$$\frac{\partial p^*}{\partial r^*} - \rho_{nf} \frac{\acute{u}^2}{Y} = 0 \tag{2}$$

$$\acute{v} \frac{\partial \acute{u}}{\partial r^*} + \frac{\acute{u}}{Y} \frac{\partial \acute{u}}{\partial s^*} + \frac{\acute{u}\acute{v}}{Y} = -\frac{1}{\rho_{nf}} \frac{\mathcal{R}^*}{Y} \frac{\partial p^*}{\partial s^*} + \frac{\mu_{nf}}{\rho_{nf}} \left\{ \frac{\partial^2 \acute{u}}{\partial r^{*2}} + \frac{1}{Y} \frac{\partial \acute{u}}{\partial r^*} - \frac{\acute{u}}{Y^2} \right\} \tag{3}$$

$$\acute{v} \frac{\partial \mathcal{T}^*}{\partial r^*} + \frac{\mathcal{R}^* \acute{u}}{Y} \frac{\partial \mathcal{T}^*}{\partial s^*} = \frac{k_{nf}}{(\rho C_p)_{nf}} \left[\frac{\partial^2 \mathcal{T}^*}{\partial r^{*2}} + \frac{1}{Y} \frac{\partial \mathcal{T}^*}{\partial r^*} - \frac{1}{k_{nf}Y} \frac{\partial}{\partial r^*}(Y)\tilde{q}_{r^*} + \frac{\acute{Q}}{k_{nf}}(\mathcal{T}^* - \mathcal{T}^*_\infty) \right] \tag{4}$$

where $Y = r^* + \mathcal{R}^*$. In the above set of equations, the velocities along s^* and r^* are denoted by the symbols $\acute{u}$ and $\acute{v}$, respectively. Moreover, p^* represents the dimensional pressure, $\tilde{q}_{r^*}$ describes the radiative heat flux and $\mathcal{T}^*$ is the base fluid temperature. The term $\acute{Q}$ arises due to heat generation and it represents the volumetric rate of heat generation through a source. Furthermore, $(\rho C_p)_{nf}$ denotes volumetric heat capacity of the nanofluid and μ_{nf} is its dynamic viscosity. Effective density and the thermal conductivity of the nanofluid are respectively symbolized by ρ_{nf}, and k_{nf} [11,27,49]. Mathematically:

$$v_{nf} = \left(\mu\rho^{-1}\right)_{nf}, \quad \mu_{nf} = \mu_f(1-\Phi)^{-5/2}, \quad \alpha_{nf} = \left(\frac{k}{\rho C_p}\right)_{nf}, \quad \frac{\rho_{nf}}{\rho_f} = \left[1-\left(1-\frac{\rho_{CNT}}{\rho_f}\right)\Phi\right] \quad (5)$$

$$\frac{(\rho C_p)_{nf}}{(\rho C_p)_f} = \left[1-\left(1-\frac{(\rho C_p)_{CNT}}{(\rho C_p)_f}\right)\Phi\right] \quad (6)$$

$$\frac{k_{nf}}{k_f} = \left(\frac{(1-\Phi)\left(k_{CNT}-k_f\right)+2\Phi k_{CNT}\ln\left(\frac{k_{CNT}+k_f}{2k_f}\right)}{(1-\Phi)\left(k_{CNT}-k_f\right)+2\Phi k_f\ln\left(\frac{k_{CNT}+k_f}{2k_f}\right)}\right) \quad (7)$$

where the base fluid viscosity is represented by μ_f, Φ denotes the nanoparticle volume fraction and α_{nf} is the thermal diffusivity. Moreover, $\left(k_f, k_{CNT}\right)$ and $\left(\rho_f, \rho_{CNT}\right)$ represent the thermal conductivities and densities of base fluid and CNT, respectively. The following table highlights the thermo-physical properties of pure water (base fluid) and CNTs.

The associated boundary conditions are imposed as follows:

$$\text{At } r^*=0, \quad \acute{u}=\acute{u}_w=\acute{a}s^*, \quad \acute{v}=0, \quad T^*=T^*{}_w, \quad \acute{u}\to 0, \quad \frac{\partial\acute{u}}{\partial r^*}\to 0, \quad T^*\to T^*{}_\infty \text{ as } r^*\to\infty \quad (8)$$

The radiative heat flux is defined by the means of Rosseland approximation [50] as:

$$\tilde{q}_{r^*} = -\frac{16\acute{\sigma}T^{*3}}{3a_R}\frac{\partial T^{*4}}{\partial r^*} \quad (9)$$

where Stefan–Boltzmann constant and Rosseland mean absorption coefficient, are $\acute{\sigma}$ and a_R, respectively. The small changes in temperature diffusion are assumed in such a way that we may expand the Taylor series expansion of T^{*4} about $T^*{}_\infty$ and ignore the terms of higher order. Consequently, we get

$$T^{*4} \cong -\left(3T^{*4}_\infty - 4T^{*3}_\infty T^*\right) \quad (10)$$

By substituting Equations Equations (9) and (10) into Equation (4), we get

$$\acute{v}\frac{\partial T^*}{\partial r^*} + \frac{R^*\acute{u}}{Y}\frac{\partial T^*}{\partial s^*} = \frac{k_{nf}}{(\rho C_p)_{nf}}\left(1+\frac{16\acute{\sigma}T^{*3}_\infty}{3a_R k_f(k_{nf}/k_f)}\right)\left[\frac{\partial^2 T^*}{\partial r^{*2}} + \frac{1}{Y}\frac{\partial T^*}{\partial r^*}\right] + \frac{\acute{Q}}{(\rho C_p)_{nf}}(T^*-T^*{}_\infty) \quad (11)$$

Now by assuming $Rd = 16\acute{\sigma}T^{*3}_\infty/3a_R k_f$ as a radiation parameter, following Magyari and Pantokratoras [51], Equation (11) becomes

$$\acute{v}\frac{\partial T^*}{\partial r^*} + \frac{R^*\acute{u}}{Y}\frac{\partial T^*}{\partial s^*} = \frac{Y_f}{\Lambda_2}\frac{1}{Pr}\frac{k_{nf}}{k_f}\left(1+\frac{Rd}{(k_{nf}/k_f)}\right)\left[\frac{\partial^2 T^*}{\partial r^{*2}} + \frac{1}{Y}\frac{\partial T^*}{\partial r^*}\right] + \frac{\acute{Q}}{(\rho C_p)_{nf}}(T^*-T^*{}_\infty) \quad (12)$$

where Prandtl number is given as $Pr = v_f/\alpha_f$.

The dimensionless transformation variables, for the simplification of the flow equations, are defined as:

$$\acute{u}=\acute{a}s^*f'(\zeta), \quad \acute{v}=-\frac{R^*}{Y}\sqrt{\acute{a}v_f}f(\zeta), \quad \zeta=\sqrt{\frac{\acute{a}}{v_f}}r^*, \quad \acute{p}=\rho\acute{a}^2s^{*2}P(\zeta), \quad T^*=T^*{}_\infty+\frac{As^*}{\ell}\Theta(\zeta), \quad \Theta(\zeta)=\frac{T^*-T^*{}_\infty}{T^*{}_w-T^*{}_\infty} \quad (13)$$

The substitution of above equation implies an automatic satisfaction of the continuity Equation (1), while Equations (2), (3) and (12) are given as:

$$\frac{\partial P}{\partial \zeta} = \frac{f'^2}{\zeta + \kappa} \tag{14}$$

$$\frac{2\kappa}{\zeta + \kappa} P = \frac{v_{nf}}{v_f} \left[f''' + \frac{f''}{\zeta + \kappa} - \frac{f'}{(\zeta + \kappa)^2} \right] - \frac{\kappa}{\zeta + \kappa} f'^2 + \frac{\kappa}{\zeta + \kappa} f f'' + \frac{\kappa}{(\zeta + \kappa)^2} f f' \tag{15}$$

$$\left(1 + \frac{Rd}{\left(k_{nf}/k_f \right)} \right) \left(\Theta'' + \frac{\Theta\prime}{\zeta + \kappa} \right) - \frac{Pr}{\left(k_{nf}/k_f \right)} \left[\Lambda_2 \frac{\kappa}{\zeta + \kappa} \left(f'\Theta - f\Theta\prime \right) - \lambda\Theta \right] = 0 \tag{16}$$

where dimensionless radius of curvature and heat generation parameter are $\kappa = \mathcal{R}^* \sqrt{\acute{a}/Y_f}$ and $\lambda = \acute{Q}/\acute{a}(\rho C_p)_f$, respectively. The classical energy equation can be recovered from Equation (6) by ignoring the radiative ($Rd = 0$) and heat generation ($\lambda = 0$) effects. Moreover, it is noteworthy that by taking $\kappa \to \infty$ and in the absence of pressure gradient, Equation (15) gets converted to the classical problem of flat stretching sheet as discussed by Crane [30],

$$f''' - f'^2 + f f'' = 0.$$

The implementation of Equation (13) reduces the boundary conditions to a dimensionless form given as:

$$f(0) = 0, \quad f'(0) = 1, \quad \Theta(0) = 1, f'(\infty) = 0, \quad f''(\infty) = 0, \quad \Theta(\infty) = 0 \tag{17}$$

We can eliminate P from Equations (14) and (15). Consequently,

$$f^{iv} + \frac{2f'''}{\zeta + \kappa} - \frac{f''}{(\zeta + \kappa)^2} + \frac{f'}{(\zeta + \kappa)^3} + \Lambda_1 \left[\begin{array}{c} \frac{\kappa}{\zeta + \kappa}(ff''' - f'f'') \\ -\frac{\kappa}{(\zeta+\kappa)^2}\left(f'^2 - ff'' \right) - \frac{\kappa}{(\zeta+\kappa)^3} ff' \end{array} \right] = 0 \tag{18}$$

where

$$\left. \begin{array}{c} \Lambda_1 = \frac{v_f}{v_{nf}} = (1 - \Phi)^{5/2} \left[1 - \left(1 - \frac{\rho_{CNT}}{\rho_f} \right) \Phi \right], \\ \Lambda_2 = \left[1 - \left(1 - \frac{(\rho C_p)_{CNT}}{(\rho C_p)_f} \right) \Phi \right] \end{array} \right\} \tag{19}$$

After obtaining $f(\zeta)$, $P(\zeta)$ can be easily determined from Equation (15) as:

$$P(\zeta) = \frac{\zeta + \kappa}{2\kappa} \left[\frac{1}{\Lambda_1} \left\{ f''' + \frac{f''}{\zeta + \kappa} - \frac{f'}{(\zeta + \kappa)^2} \right\} - \frac{\kappa}{\zeta + \kappa} f'^2 + \frac{\kappa}{\zeta + \kappa} f f'' + \frac{\kappa}{(\zeta + \kappa)^2} f f' \right] \tag{20}$$

The skin-friction coefficient C_f and the local Nusselt number Nu_{s^*} in s^*-direction are given by:

$$C_f = \frac{(\tau_{r^*s^*})_{r^*=0}}{\rho_f \acute{u}_w^2}, \quad Nu_{s^*} = \frac{s^* \tilde{q}_w/k_f}{(T^*_w - T^*_\infty)} \tag{21}$$

where $\tau_{r^*s^*}$ represents the wall shear stress and $\tilde{q}_w$ symbolizes the wall heat flux in s^*-direction. Their mathematical expression are as follows:

$$\tau_{r^*s^*} = \mu_{nf} \left(\frac{\partial \acute{u}}{\partial r^*} - \frac{\acute{u}}{Y} \right)\Big|_{r^*=0}, \quad \tilde{q}_w = -k_{nf} \left(1 + \frac{16\acute{\sigma}T^{*3}_\infty}{3a_R k_f \left(k_{nf}/k_f \right)} \right) \frac{\partial T^*}{\partial r^*}\Big|_{r^*=0} \tag{22}$$

Using Equations (13), (21) and (22) becomes:

$$Re_{s^*}^{1/2}C_f = \frac{1}{(1-\Phi)^{2.5}}\left(f''(0) - \frac{f'(0)}{\kappa}\right), \quad Re_{s^*}^{-1/2}Nu_{s^*} = -\frac{k_{nf}}{k_f}\left(1 + \frac{Rd}{\left(k_{nf}/k_f\right)}\right)\Theta'(0) \qquad (23)$$

where the local Reynolds number is symbolized by $Re_{s^*} = \acute{a}s^{*2}/v_f t$.

3. Solution Procedure

The shooting method, followed by Runge–Kutta–Fehlberg scheme, has been implemented to solve the problem. The shooting method aids in transforming the given system into an initial value problem which can then be solved by Runge–Kutta–Fehlberg (RKF) method. Mathematical software Mathematica (Version 10, Wolfram: Computation Meets Knowledge, Champaign, IL, USA) has been used to solve the present problem by assuming the step size $\Delta\zeta = 0.001$, and the convergence criteria is fixed at 10^{-6}.

4. Results and Discussions

The size and shape of nanocomposites matter greatly for enhaning the thermal properties of the base fluid. Carbon nanotubes are long nanotube wires with distinctive size and shape and also possess significant physical properties. They are supposed to be much better than steel and Kevlar due to the material's exceptional tensile strength and stiffness. Thus, the flow and heat transfer characteristics of single- and multi-walled carbon nanotubes over a curved stretching sheet have been investigated. The mass flow rate is assumed to be 0.02 kg/s. The core objective of this section is to graphically demonstrate the influence of various parameters like dimensionless radius of curvature κ, nanoparticle volume fraction Φ (ranging from $0 \leq \Phi \leq 0.2$), heat generation parameter λ and radiation parameter Rd on the velocity, temperature and pressure profiles. Table 1 displays the thermo-physical properties of the base fluid and the nanoparticles. Since water has been utilized as a base fluid, therefore, a fixed value for Prandtl number ($Pr = 6.2$) has been used.

Table 1. Physical and thermal properties of pure water and CNTs [11,27]. CNTs: carbon nanotubes; SWCNTs: Single-walled carbon nanotubes; MWCNTs: multi-walled carbon nanotubes.

	Pure Water	**SWCNTs**	**MWCNTs**
C_p (J(kgK)$^{-1}$)	4179	425	796
$\acute{\rho}$ (kg(m)$^{-3}$)	997.1	2600	1600
k (W(mK)$^{-1}$)	0.613	6600	3000

Figure 2 visualizes the effects of dimensionless curvature on the velocity, temperature and pressure profiles. Figure 2a depicts the graphical description of the velocity profile for both SWCNTs and MWCNTs based nanofluid under the influence of dimensionless radius of curvature κ. One can clearly notice an increase in the velocity as well as the momentum boundary layer thickness, with increasing dimensionless curvature (i.e., decreasing κ). As the fluid particles trace the curved path along the surface of the sheet, the curvature of the sheet, under the influence of centrifugal force, enhances the secondary flow of the fluid. This flow is relatively minor, which is then superimposed on the primary flow which as a result enhances the fluid velocity.

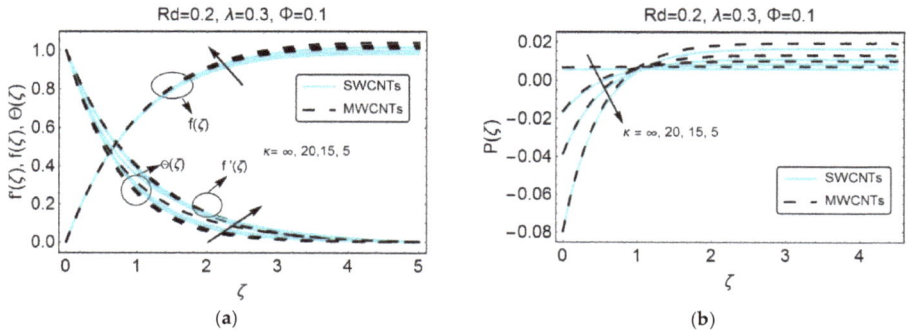

Figure 2. (a) $f(\zeta)$, $f'(\zeta)$ and $\Theta(\zeta)$; and (b) $P(\zeta)$ for some values of κ.

The effect of curvature seems to be small in the s^*-direction, while, in the r^*-direction, the impact is quite prominent which is mainly due to the centrifugal force acting towards the origin. Figure 2a also exhibits the role of dimensionless radius of curvature κ on temperature profile $\Theta(\zeta)$. The increment of the dimensionless curvature enhances the temperature of the fluid, i.e., the maximum amount of heat is transfer when the radius of curvature decreases. Figure 2b illustrates the influence of dimensionless curvature on the pressure profile. Clearly, a decline in dimensionless radius of curvature enhances the magnitude of pressure inside the boundary layer. However, far away from the boundary, the pressure remains almost negligible because the behavior of stream lines away from the sheet follows the same pattern of flow as in the case of flat stretching sheet. Furthermore, it is pertinent to mention that the pressure variations, near and away from the surface, seem insignificant in the case of flat stretching sheet (i.e., $\kappa = 1000$). On the other hand, the pressure does not remain constant when the curvature is employed into the sheet and, therefore, a certain variation has been perceived, especially inside the boundary layer.

Figure 3 has been placed to show the influence of nanoparticle volume fraction Φ on the velocity, temperature and pressure profiles. A certain rise in velocity has been observed when the nanoparticle volume fraction increases. Besides, it has been observed that the SWCNTs have a slightly lower velocity than MWCNTs, which is due to the higher density values for SWCNTs with which the resistance within the fluid increases and consequently the dimensionless horizontal velocity component, i.e., $f'(\zeta)$, for SWCNTs decreases. Figure 3a also shows the variations in temperature $\Theta(\zeta)$ with increasing Φ. The temperature of the fluid rises and consequently, the thermal boundary layer increases. Since the carbon nanotubes bear high thermal conductivity and low specific heat as compared to the base fluid, the inclusion of sufficient nanoparticles enhances the temperature of the fluid quite significantly. Moreover, the temperature for MWCNT-nanoparticles exhibits lower values as compared to SWCNT-nanoparticles. Figure 3b depicts the behavior of pressure distribution for distinct values of solid volume fraction Φ. A decline has been noticed in the magnitude of pressure distribution with an increasing Φ.

The results in Figure 4 highlight the variations in temperature for increasing values of radiative factor Rd. As expected, because of radiation, nanofluid temperature raises quite significantly. The reason behind is the decline in mean absorption coefficient and as a result the rate with which radiative heat transfers seems to be increasing at every point distant from the sheet. It has also been observed that the temperature of SWCNTs-nanoparticles increases more than MWCNTs nanoparticles because of the less thermal conductivity of the MWCNTs nanoparticles. Figure 5 depicts the behavior of heat generation parameter λ on the temperature profile. The increasing $\lambda > 0$ causes an upsurge in temperature as well as in thermal boundary layer thickness for both SWCNTs and MWCNTs nanoparticle. The heat generation parameter usually involves the heat generation coefficient and, thus, the increment of heat generation parameters leads to an increase in heat generation coefficient which means that the temperature at the surface is higher than the free stream temperature and, therefore, by transferring the heat from the sheet to the fluid, the rise in the temperature is quite obvious. Moreover, the increment of heat absorption parameter $\lambda < 0$ implies a decline in temperature in the region adjacent to the wall for both SWCNTs and MWCNTs based nanofluid.

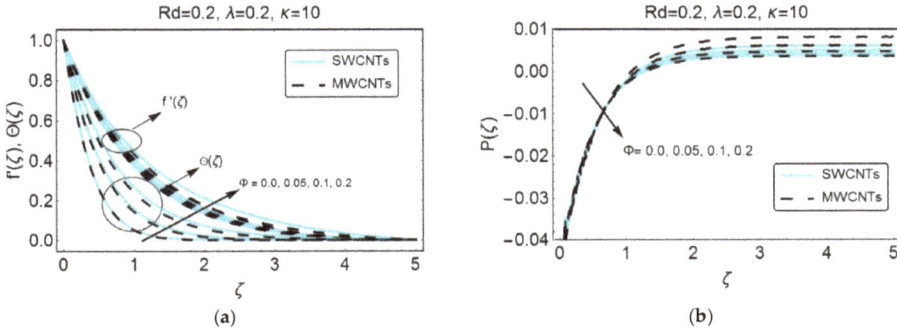

Figure 3. (**a**) $f'(\zeta)$ and $\Theta(\zeta)$; and (**b**) $P(\zeta)$ for some values of Φ.

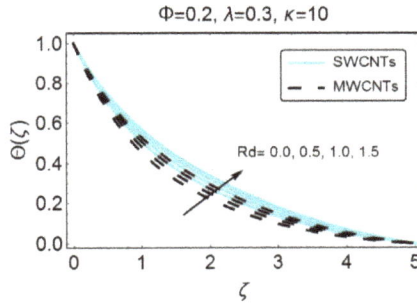

Figure 4. $\Theta(\zeta)$ for some values of Rd.

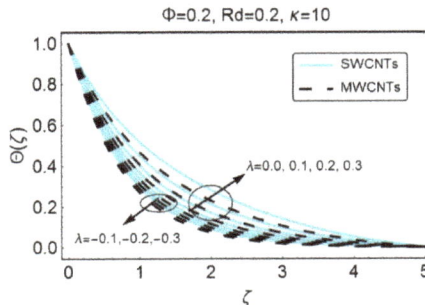

Figure 5. $\Theta(\zeta)$ for some values of λ.

Figure 6 describes the influential behavior of various parameters on skin friction coefficient. The increment in nanoparticle volume fraction considerably enhances the magnitude of skin friction coefficient, as seen in Figure 6a. The reason is the higher volume fraction of nanoparticles, in terms of composite carbon chains, which plays a prominent role within the fluid as well as on the surface and as a result increases the surface friction in both directions. Furthermore, it has been observed that the SWCNTs remain on the higher side because the higher density value of SWCNTs as compared to MWCNTs, therefore, the resistance within the fluid, seems to be more significant in the case of SWCNTs and thus exhibits the higher skin friction. Figure 6b depicts the impact of dimensionless radius of curvature κ on skin friction coefficient for both SW- and MW-carbon nanotubes. A decline in skin friction coefficient has been observed with the increasing values of dimensionless radius of curvature κ, which indicates that the increment in κ decreases the curvature of the surface and, therefore, offers a less resistance to the fluid particles which eventually decay the skin friction coefficient.

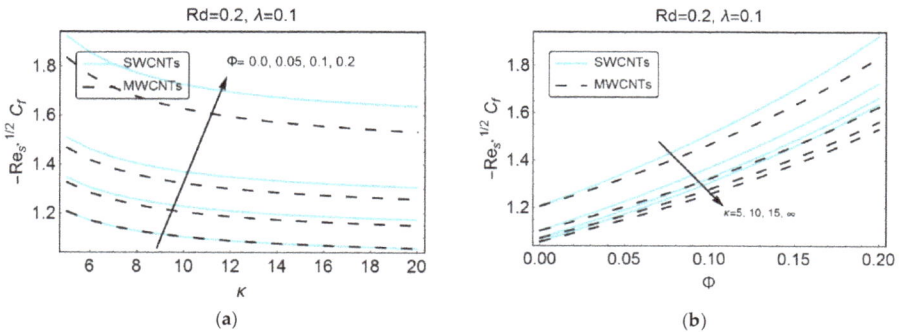

Figure 6. Skin friction coefficient for some values of: (**a**) Φ; and (**b**) κ.

Figure 7 illustrates the effects of various parameters such as solid volume fraction Φ, heat generation parameter λ, radiation parameter Rd and dimensionless radius of curvature κ. on the rate of heat transfer for both single walled and multi walled carbon nanotubes. In Figure 7a, one can clearly observe an enhancement in the local heat flux rate with an increasing nanoparticle volume fraction Φ for both SW- and MW-carbon nanotubes. Nusselt number is usually a product of thermal conductivity ratio and temperature gradient.The temperature gradient experiences a decline due to the inclusion of nanoparticles and it is much smaller than the thermal conductivity ratio, which as a result enhances the magnitude of Nusselt number. Furthermore, it has also been noticed in Figure 7a that the local heat flux experiences a decline with the increasing heat generation parameter $\lambda > 0$. Since the increment of heat generation parameter significantly raises the temperature of the fluid, the decrement in the rate of heat transfer is obvious. The impact of radiation parameter Rd on the magnitude of the local Nusselt number is presented in Figure 7b. It has been noticed that the rate of heat flux is an increasing function of Rd. The increment of radiation parameter Rd drastically enhances the heat transfer coefficient which as a result shows an increment in heat transfer rate. Moreover, Figure 7b also describes the effect of dimensionless radius of curvature κ on the rate of heat flux. The increment in the magnitude of local Nusselt number has been perceived with an increasing dimensionless curvature (i.e., decreasing κ). Since the temperature of the fluid rises with an increasing dimensionless curvature, rate of heat flux seems to be significant with decreasing κ.

Table 2 has been organized to see the variations in thermal and physical properties of base fluid with the dispersion of carbon nanotubes. It has been clearly noticed that the inclusion of both types of carbon nanotubes significantly enhances the thermal conductivity of the base fluid. The density of the fluid also increases while the specific heat experiences a clear decline. Tables 3 and 4 has been organized to discover the impact of various embedded parameters on local skin friction and local Nusselt number for both SW- and MW-carbon nanotubes. Table 3 illustrates the influence of both solid volume fraction Φ and dimensionless radius of curvature κ on skin friction coefficient C_f. In addition, comparison with the previously existing results of Abbas et al. [44] has been presented. It has been witnessed that the skin friction coefficient experiences a decline because of increasing dimensionless radius of curvature κ which indicates that a larger drag force is required to drag the fluid over the curved stretching sheet. Table 4 provides the numerical values for skin friction coefficient $-Re_{s*}^{1/2}C_f$ and the magnitude of local Nusselt number $Re_{s*}^{-1/2}Nu_{s*}$ for numerous values of solid volume fraction Φ and heat generation parameter λ under the assumption that the dimensionless radius of curvature $\kappa \rightarrow \infty$ i.e., $\kappa = 1000$ (for flat dilating sheet). It has been observed that the skin friction coefficient increases with increasing values of Φ, which can also be observed in Table 3. Moreover, less resistance has been offered by MWCNTs as compared to SWCNTs. In addition, the magnitude of local Nusselt number seems to be an increasing function of Φ for both SWCNTs and MWCNTs. The reason is the supremacy of CNTs over traditional base fluid, i.e., water, which is primarily due to the higher thermal conductivity and low specific heat. Thus, the ability to conduct heat significantly improves the rate of heat flux at the sheet. This proves the fact that the addition of carbon nanotubes enhances the local heat flux. On the other hand, the local rate of heat flux experiences a decrement with the increasing values of λ for CNTs based nanofluid as well as traditional base fluid. Finally, the validity of our results for a dimensionless radius of curvature $\kappa \rightarrow \infty$, i.e., $\kappa = 1000$ (for flat dilating sheet), is shown in Table 5. An excellent agreement has been found between obtained results and previously published results of Ali [52], Grubka and Bobba [53] and Ishak et al. [54] in the absence of radiation parameter Rd, heat generation parameter λ and carbon nanotubes Φ.

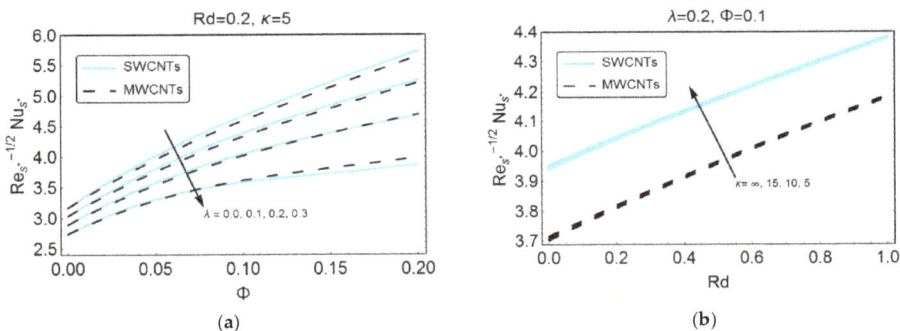

Figure 7. Nusselt number for some values of (a) λ; (b) κ.

Table 2. Variation in Thermo-physical properties of CNTs based nanofluid.

	Φ	ρ	$\rho C_p \times 10^6$	k
SWCNT	0.0	997.1	4.1669	0.613
	0.05	1077.25	4.0138	1.1673
	0.10	1157.39	3.8607	1.7831
	0.15	1237.54	3.7076	2.4712
	0.20	1317.68	3.5545	3.2451
MWCNT	0.0	997.1	4.1669	0.613
	0.05	1027.25	4.0222	1.1164
	0.10	1057.39	3.8776	1.6755
	0.15	1087.54	3.7329	2.3002
	0.20	1117.68	3.5882	3.0027

Table 3. Comparison of $-Re_{s^*}^{1/2}C_f$ between present results and previously existing results of Abbas et al. [44] for various values of Φ and κ.

Φ	κ	Abbas et al. [44]	SWCNT	MWCNT
0.0	5	1.15763	1.15763	1.15763
	10	1.07349	1.07349	1.07349
	20	1.03561	1.03561	1.03561
	50	1.01405	1.01405	1.01405
	100	1.00704	1.00704	1.00704
	200	1.00356	1.00356	1.00356
	1000	1.00079	1.00079	1.00079
	∞	1.00000	1.00000	1.00000
0.1	5	-	1.43781	1.38677
	10	-	1.32577	1.27251
	20	-	1.27579	1.22190
0.2	5	-	1.80802	1.69810
	10	-	1.65195	1.53544
	20	-	1.58323	1.46471

Table 4. Comparison of $-Re_{s^*}^{1/2}C_f$ and $Re_{s^*}^{-1/2}Nu_{s^*}$ for different values of Φ and λ by setting $Rd = 0$ and $\kappa = 1000$ fixed.

	Φ	$\lambda = -0.3$		$\lambda = 0.0$		$\lambda = 0.3$	
		$-Re_{s^*}^{1/2}C_f$	$Re_{s^*}^{-1/2}Nu_{s^*}$	$-Re_{s^*}^{1/2}C_f$	$Re_{s^*}^{-1/2}Nu_{s^*}$	$-Re_{s^*}^{1/2}C_f$	$Re_{s^*}^{-1/2}Nu_{s^*}$
SWCNT	0.0	1.00001	3.19358	1.00001	2.87661	1.00001	2.51646
	0.1	1.22906	5.06371	1.22906	4.43995	1.22906	3.66480
	0.2	1.51945	6.42309	1.51945	5.47268	1.51945	4.01986
MWCNT	0.0	1.00001	3.19358	1.00001	2.87661	1.00001	2.51646
	0.1	1.17477	4.95614	1.17477	4.36549	1.17477	3.64727
	0.2	1.39944	6.29734	1.39944	5.42743	1.39944	4.25578

Table 5. Comparison of $-\Theta'(0)$ between present results and previously existing results for different values of Pr by setting $Rd = \lambda = \Phi = 0$ and $\kappa \to \infty$.

Pr	Ali [52]	Grubka and Bobba [53]	Ishak et al. [54]	Present Results
0.72	0.8058	0.8086	0.8086	0.80884
1	1.0000	1.0000	1.0000	1.00001
3	1.9237	1.9144	1.9237	1.92368
7	-	-	3.0723	3.07226
10	3.7006	3.7207	3.7207	3.72068
100	-	12.2940	12.2941	12.29407

5. Conclusions

A time independent, steady, incompressible boundary layer flow of CNT (single-walled and multi-walled carbon nanotubes) based nanofluid over a curved dilating sheet has been considered. Impact of different physical parameters have enabled us to draw the following conclusions:

- The increment in dimensionless curvature (i.e., decreasing κ) and nanoparticle volume fraction Φ significantly enhances the fluid velocity. Moreover, velocity dominates for multi-walled carbon nanotubes.
- An increment in temperature along with thermal boundary layer has been observed for increasing dimensionless curvature (i.e., decreasing κ), nanoparticle volume fraction Φ and radiation parameter Rd. Furthermore, SWCNTs remains on the higher side as compared to MWCNTs.

Appl. Sci. **2018**, *8*, 395

- An upsurge in temperature has been perceived for increasing heat generation parameter $\lambda > 0$ while the opposite behavior has been detected for increasing heat absorption parameter $\lambda < 0$.
- A decline is observed in the magnitude of pressure distribution inside the boundary layer as the dimensionless radius of curvature increases while a rise in nanoparticle volume fraction depicts a clear enhancement in it.
- Skin friction enhances for higher values of solid volume fraction Φ while dimensionless radius of curvature κ implies a reduction in skin friction coefficient for both SWCNTs and MWCNTs.
- The magnitude of local heat flux rate increases for growing values of dimensionless curvature (i.e., decreasing κ), nanoparticle volume fraction Φ and radiation parameter Rd while the behavior is opposite for increasing heat generation parameter λ.

Acknowledgments: The work here is supported by grant: UKM-2017-064. Moreover, first author Syed Tauseef Mohyud-Din is also thankful to Chairman Bahria Town/ Patron and Chairman of FAIRE; Chief Executive FAIRE; Administration of University of Islamabad (a project of Bahria Town) for establishment of Center for Research (CFR) and the provision of conducive research environment.

Author Contributions: Fitnat Saba and Naveed Ahmed developed the problem and its MATHEMATICA code. Umar Khan in collaboration with Syed Tauseef Mohyud-Din, did the literature review, developed and implemented the computer code, and interpreted the subsequently obtained results. Saqib Hussain did the editing and removed the grammatical mistakes. Maslina Darus in consultation of rest of the Authors, re-confirmed the credibility of obtained solutions. All authors have read and approved the final manuscript.

Conflicts of Interest: The authors declare no conflict of interest.

Nomenclature

$\acute{A}$	Constant
κ	Dimensionless radius of curvature
k	Thermal conductivity, W/mK
Rd	Radiation parameter
C_p	Specific heat at constant pressure, J/kg·K
$\acute{p}$	Pressure
P	Dimensionless pressure
Pr	Prandtl number
$\acute{Q}$	Volumetric rate of heat source
$\tilde{q}_{r^*}$	Radiative heat flux, W/m^2
$\tilde{q}_w$	Wall heat flux, W/m^2
$\mathcal{R}^*$	Radius of curvature
Re_{s^*}	Local Reynold number
Nu_{s^*}	Local Nusselt number
T^*	Local fluid temperature, K
$T^*{}_w$	Surface temperature, K
$T^*{}_\infty$	Free stream temperature, K
$\acute{u}$	s^*-component of velocity, m/s
$\acute{v}$	r^*-component of velocity, m/s

Greek Symbols

Φ	Solid volume fraction of CNT
ζ	Similarity variable
μ	Dynamic viscosity, N·s/m^2
v	Kinematic viscosity, m^2/s
$\acute{\rho}$	Density, kg/m^3
α	Thermal diffusivity, m^2/s
$\acute{\rho}C_p$	Heat capacitance
Θ	Dimensionless temperature
λ	Heat generation parameter

$\tau_{r^*s^*}$	Wall shear stress
$\acute{\sigma}$	Stefan–Boltzmann constant
a_R	Mean absorption coefficient

Subscripts

nf	Nanofluid
f	Base fluid
CNT	Carbon nanotube
SW	Single walled
MW	Multi walled

References

1. Nnanna, A.G.A. Experimental model of temperature-driven nanofluids. *J. Heat Transf.* **2007**, *129*, 697–704. [CrossRef]
2. Gu, B.; Hou, B.; Lu, Z.; Wang, Z.; Chen, S. Thermal conductivity of nanofluids containing high aspect ratio fillers. *Int. J. Heat Mass Transf.* **2013**, *64*, 108–114. [CrossRef]
3. Milanese, M.; Iacobazzi, F.; Colangelo, G.; De Risi, A. An investigation of layering phenomenon at the liquid-solid interface in Cu and CuO based nanofluids. *Int. J. Heat Mass Transf.* **2016**, *103*, 564–571. [CrossRef]
4. Iacobazzi, F.; Milanese, M.; Colangelo, G.; Lomascolo, M.; De Risi, A. An explanation of the Al_2O_3 nanofluid thermal conductivity based on the phonon theory of liquid. *Energy* **2016**, *116*, 786–794. [CrossRef]
5. Colangelo, G.; Milanese, M.; De Risi, A. Numerical simulation of thermal efficiency of an innovative Al_2O_3 nanofluid solar thermal collector: Influence of nanoparticles concentration. *Therm. Sci.* **2016**, *2016*, 2769–2779. [CrossRef]
6. Choi, S.U.S. Enhancing thermal conductivity of fluids with nanoparticles. *Dev. Appl. Non Newton. Flows* **1995**, *231*, 99–105.
7. Choi, S.U.S.; Zhang, Z.G.; Yu, W.; Lockwood, F.E.; Grulke, E.A. Anomalous thermal conductivity enhancement in nanotube suspensions. *Appl. Phys. Lett.* **2001**, *79*, 2252–2254. [CrossRef]
8. Buongiorno, J. Convective transport in nanofluids. *J. Heat Transf.* **2006**, *128*, 240–250. [CrossRef]
9. Hamilton, R.L.; Crosser, O.K. Thermal conductivity of heterogeneous two-component systems. *Ind. Eng. Chem. Fundam.* **1962**, *1*, 187–191. [CrossRef]
10. Maxwell, J.C. *A Treatise on Electricity and Magnetism*, 3rd ed.; Oxford University Press: Oxford, UK, 1904.
11. Xue, Q.Z. Model for thermal conductivity of carbon nanotube-based composites. *Phys. B Condens. Matter* **2005**, *368*, 302–307. [CrossRef]
12. Khan, W.A.; Pop, I. Boundary-layer flow of a nanofluid past a stretching sheet. *Int. J. Heat Mass Transf.* **2010**, *53*, 2477–2483. [CrossRef]
13. Kandelousi, M.S.; Ellahi, R. Simulation of Ferrofluid Flow for Magnetic Drug Targeting Using the Lattice Boltzmann Method. *J. Z. Naturforschung A* **2015**, *70*, 115–124. [CrossRef]
14. Akbar, N.S.; Raza, M.; Ellahi, R. Influence of induced magnetic field and heat flux with the suspension of carbon nanotubes for the peristaltic flow in a permeable channel. *J. Magn. Magn. Mater.* **2015**, *381*, 405–415. [CrossRef]
15. Sheikholeslami, M.; Ganji, D.D.; Younus Javed, M.; Ellahi, R. Effect of thermal radiation on magnetohydrodynamics nanofluid flow and heat transfer by means of two phase model. *J. Magn. Magn. Mater.* **2015**, *374*, 36–43. [CrossRef]
16. Rashidi, S.; Dehghan, M.; Ellahi, R.; Riaz, M.; Jamal-Abad, M.T. Study of stream wise transverse magnetic fluid flow with heat transfer around an obstacle embedded in a porous medium. *J. Magn. Magn. Mater.* **2015**, *378*, 128–137. [CrossRef]
17. Mabood, F.; Khan, W.A.; Ismail, A.I.M. MHD boundary layer flow and heat transfer of nanofluids over a nonlinear stretching sheet: A numerical study. *J. Magn. Magn. Mater.* **2015**, *374*, 569–576. [CrossRef]
18. Mohyud-Din, S.T.; Zaidi, Z.A.; Khan, U.; Ahmed, N. On heat and mass transfer analysis for the flow of a nanofluid between rotating parallel plates. *Aerosp. Sci. Technol.* **2015**, *46*, 514–522. [CrossRef]
19. Mohyud-Din, S.T.; Khan, U.; Ahmed, N.; Hassan, S. Magnetohydrodynamic Flow and Heat Transfer of Nanofluids in Stretchable Convergent/Divergent Channels. *Appl. Sci.* **2015**, *5*, 1639–1664. [CrossRef]
20. Khan, U.; Ahmed, N.; Mohyud-Din, S.T. Thermo-diffusion, diffusion-thermo and chemical reaction effects on MHD flow of viscous fluid in divergent and convergent channels. *Chem. Eng. Sci.* **2016**, *141*, 17–27. [CrossRef]

21. Iijima, S. Helical microtubules of graphitic carbon. *Nature* **1991**, *354*, 56–58. [CrossRef]
22. Saito, R.; Dresselhaus, G.; Dresselhaus, M.S. *Physical Properties of Carbon Nanotubes*; Imperial College Press: Singapore, 2001.
23. Endo, M.; Hayashi, T.; Kim, Y.A.; Terrones, M.; Dresselhaus, M.S. Applications of carbon nanotubes in the twenty–first century. *Philos. Trans. R. Soc. Lond. Ser. A Math. Phys. Eng. Sci.* **2004**, *362*, 2223–2238. [CrossRef] [PubMed]
24. Murshed, S.M.S.; Nieto de Castro, C.A.; Lourenço, M.J.V.; Lopes, M.L.M.; Santos, F.J.V. A review of boiling and convective heat transfer with nanofluids. *Renew. Sustain. Energy Rev.* **2011**, *15*, 2342–2354. [CrossRef]
25. Akbar, N.S.; Butt, A.W. CNT suspended nanofluid analysis in a flexible tube with ciliated walls. *Eur. Phys. J. Plus* **2014**, *129*, 174. [CrossRef]
26. Ul Haq, R.; Nadeem, S.; Khan, Z.H.; Noor, N.F.M. Convective heat transfer in MHD slip flow over a stretching surface in the presence of carbon nanotubes. *Phys. B Condens. Matter* **2015**, *457*, 40–47. [CrossRef]
27. Ahmed, N.; Mohyud-Din, S.T.; Hassan, S.M. Flow and heat transfer of nanofluid in an asymmetric channel with expanding and contracting walls suspended by carbon nanotubes: A numerical investigation. *Aerosp. Sci. Technol.* **2016**, *48*, 53–60. [CrossRef]
28. Khan, U.; Ahmed, N.; Mohyud-Din, S.T. Heat transfer effects on carbon nanotubes suspended nanofluid flow in a channel with non-parallel walls under the effect of velocity slip boundary condition: A numerical study. *Neural Comput. Appl.* **2017**, *28*, 37–46. [CrossRef]
29. Sakiadis, B.C. Boundary-layer behavior on continuous solid surfaces: I. Boundary-layer equations for two-dimensional and axisymmetric flow. *AIChE J.* **1961**, *7*, 26–28. [CrossRef]
30. Crane, L.J. Flow past a stretching plate. *Z. Angew. Math. Phys.* **1970**, *21*, 645–647. [CrossRef]
31. McLeod, J.B.; Rajagopal, K.R. On the uniqueness of flow of a Navier-Stokes fluid due to a stretching boundary. *Arch. Ration. Mech. Anal.* **1987**, *98*, 385–393. [CrossRef]
32. Gupta, P.S.; Gupta, A.S. Heat and mass transfer on a stretching sheet with suction or blowing. *Can. J. Chem. Eng.* **1977**, *55*, 744–746. [CrossRef]
33. Brady, J.F.; Acrivos, A. Steady flow in a channel or tube with an accelerating surface velocity. An exact solution to the Navier-Stokes equations with reverse flow. *J. Fluid Mech.* **1981**, *112*, 127–150. [CrossRef]
34. Wang, C.Y. The three-dimensional flow due to a stretching flat surface. *Phys. Fluids* **1984**, *27*, 1915–1917. [CrossRef]
35. Wang, C.Y. Fluid flow due to a stretching cylinder. *Phys. Fluids* **1988**, *31*, 466–468. [CrossRef]
36. Chamkha, A.J.; Aly, A.M.; Mansour, M.A. Similarity solution for unsteady heat and mass transfer from a stretching surface embedded in a porous medium with suction/injection and chemical reaction effects. *Chem. Eng. Commun.* **2010**, *197*, 846–858. [CrossRef]
37. Bhattacharyya, K.; Uddin, M.S. Reactive Solute Diffusion in Boundary Layer Flow through a Porous Medium over a Permeable Flat Plate with Power-Law Variation in Surface Concentration. *J. Eng.* **2013**, *2013*, 1–7. [CrossRef]
38. Rashidi, M.M.; Vishnu Ganesh, N.; Abdul Hakeem, A.K.; Ganga, B. Buoyancy effect on MHD flow of nanofluid over a stretching sheet in the presence of thermal radiation. *J. Mol. Liq.* **2014**, *198*, 234–238. [CrossRef]
39. Ul Haq, R.; Nadeem, S.; Khan, Z.H.; Akbar, N.S. Thermal radiation and slip effects on MHD stagnation point flow of nanofluid over a stretching sheet. *Phys. E Low Dimens. Syst. Nanostruct.* **2015**, *65*, 17–23. [CrossRef]
40. Malvandi, A.; Hedayati, F.; Nobari, M.R.H. An HAM Analysis of Stagnation-Point Flow of a Nanofluid over a Porous Stretching Sheet with Heat Generation. *J. Appl. Fluid Mech.* **2014**, *7*, 135–145.
41. Tsai, R.; Huang, K.H.; Huang, J.S. Flow and heat transfer over an unsteady stretching surface with non-uniform heat source. *Int. Commun. Heat Mass Transf.* **2008**, *35*, 1340–1343. [CrossRef]
42. Pal, D. Combined effects of non-uniform heat source/sink and thermal radiation on heat transfer over an unsteady stretching permeable surface. *Commun. Nonlinear Sci. Numer. Simul.* **2011**, *16*, 1890–1904. [CrossRef]
43. Sajid, M.; Ali, N.; Javed, T.; Abbas, Z. Stretching a Curved Surface in a Viscous Fluid. *Chin. Phys. Lett.* **2010**, *27*, 24703. [CrossRef]
44. Abbas, Z.; Naveed, M.; Sajid, M. Heat transfer analysis for stretching flow over a curved surface with magnetic field. *J. Eng. Thermophys.* **2013**, *22*, 337–345. [CrossRef]
45. Abbas, Z.; Naveed, M.; Sajid, M. Hydromagnetic slip flow of nanofluid over a curved stretching surface with heat generation and thermal radiation. *J. Mol. Liq.* **2016**, *215*, 756–762. [CrossRef]

46. Rosca, N.C.; Pop, I. Unsteady boundary layer flow over a permeable curved stretching/shrinking surface. *Eur. J. Mech.—B/Fluids* **2015**, *51*, 61–67. [CrossRef]

47. Naveed, M.; Abbas, Z.; Sajid, M. MHD flow of Micropolar fluid due to a curved stretching sheet with thermal radiation. *J. Appl. Fluid Mech.* **2016**, *9*, 131–138. [CrossRef]

48. Hayat, T.; Rashid, M.; Imtiaz, M.; Alsaedi, A. MHD convective flow due to a curved surface with thermal radiation and chemical reaction. *J. Mol. Liq.* **2017**, *225*, 482–489. [CrossRef]

49. Khan, U.; Ahmed, N.; Mohyud-Din, S.T.; Sikander, W. Flow of carbon nanotubes suspended nanofluid in stretchable non-parallel walls. *Neural Comput. Appl.* **2017**, 1–13. [CrossRef]

50. Rosseland, S. *Astrophysik Und Atom-Theoretische Grundlagen*; Springer: Berlin, Germany, 1931.

51. Magyari, E.; Pantokratoras, A. Note on the effect of thermal radiation in the linearized Rosseland approximation on the heat transfer characteristics of various boundary layer flows. *Int. Commun. Heat Mass Transf.* **2011**, *38*, 554–556. [CrossRef]

52. Ali, M.E. Heat transfer characteristics of a continuous stretching surface. *Heat Mass Transf.* **1994**, *429*, 227–234. [CrossRef]

53. Grubka, L.J.; Bobba, K.M. Heat transfer characteristics of a continuous, stretching surface with variable temperature. *ASME J. Heat Transf.* **1985**, *107*, 248–250. [CrossRef]

54. Ishak, A.; Nazar, R.; Pop, I. Boundary layer flow and heat transfer over an unsteady stretching vertical surface. *Meccanica* **2009**, *44*, 369–375. [CrossRef]

applied
sciences

MDPI

Article

Adsorption of Nonylphenol to Multi-Walled Carbon Nanotubes: Kinetics and Isotherm Study

Yung-Dun Dai [1], Kinjal J. Shah [1,2], Ching P. Huang [3], Hyunook Kim [4,*] and Pen-Chi Chiang [1,2,*]

1   Graduate Institute of Environmental Engineering, National Taiwan University, Taipei 10637, Taiwan; d98541003@ntu.edu.tw (Y.-D.D.); d10122801@mail.ntust.edu.tw (K.J.S.)
2   Carbon Cycle Research Center, National Taiwan University, Taipei 10637, Taiwan
3   Department of Civil and Environmental Engineering, University of Delaware, Newark, DE 19716, USA; huang@udel.edu
4   Department of Environmental Engineering, University of Seoul, Seoul 02504, Korea
*   Correspondence: h_kim@uos.ac.kr (H.K.); pcchiang@ntu.edu.tw (P.-C.C.)

Received: 8 October 2018; Accepted: 7 November 2018; Published: 19 November 2018

Abstract: We explored the occurrence and distribution of nonylphenol (NP) in 13 Taiwanese source waters. From all the surveyed waters, NP was detected at a high concentration, which could be attributed to contamination by wastewater discharges. In this study, we applied modified multi-walled carbon nanotubes (MWCNTs) for removing NP from aqueous solution. The impact of a few experimental factors, i.e., pH, contact time, MWCNTs dose, and temperature on the NP removal efficiency of modified MWCNTs was studied. The maximum adsorption capacity of the MWCNTs was observed to be 1040 mg NP/g when the initial NP concentration was 2.5 mg/L, and the solution pH was 4. The adsorption process followed the Elovich kinetics and the Elovich isotherm, indicating it is multilayer adsorption. The thermodynamic analysis demonstrated the NP adsorption by MWCNTs was thermodynamically satisfactory and, for the most part, endothermic as in the case of phenol adsorption. The result of the current study demonstrated the significance of free binding sites and the pore size of MWCNTs in the NP adsorption. This paper will help to better comprehend the adsorption behavior and mechanism of alkyl phenolic compounds onto MWCNTs.

Keywords: adsorption; multi-walled carbon nanotubes; nonylphenol; kinetics

1. Introduction

Nonylphenol (NP), a hydrophobic contaminant, has frequently been found in surface water, groundwater, sediment, and soil [1,2]. It is a significant byproduct of a non-ionic surfactant, nonylphenol polyethoxylate, which is used as an ingredient of pesticides and personal care products, lubricating additive, the catalyst in curing agent of epoxy resin, defoamer in industrial laundries and dispersant in paper industries [2–4]. Nonyl phenol has a similar structure to natural estrogen, so it can mimic characteristic hormones by making a connection with estrogen receptors in the environment, which is highly toxic to fish, microorganisms, and aquatic plants. Additionally, the presence of NP in the human body will reduce sperm count and immunity and can cause breast and testicular cancers. Thus, NP has been listed as a priority substance in the EU Water-Framework Directive [5]. As a result, NP has been replaced by alcohol polyethoxylates in the most European, Canadian, and Japanese industries [1,4]. However, alcohol polyethoxylates are expensive and less efficient in general, which boosts the illegal production and usage of NP. Moreover, inadequate NP removal by wastewater treatment plants (WWTPs) results in a high NP concentration in the aquatic environment [6].

In general, the treated wastewater discharged from industrial parks in the Asian region including China, Indonesia, Korea, Laos, Malaysia, Taiwan, Thailand, and Vietnam contains a higher concentration of NP than that in the European Union, America, and Japan [6–8]. Table S1 summarizes the levels of NP detected in several major Asian rivers [6,7,9–28]. Generally, these rivers flow by industrial parks and WWTPs, which are the potential sources of NP. It is a great challenge to estimate the exact contribution by each industrial source to NP pollution of water environment, because of other significant sources such as municipal sewage, agriculture runoffs, and animal wastes [7,29]. According to the Taiwan Construction and Planning Agency Ministry of the Interior (CPAMI), only about 50% of the municipal wastewater is treated in 2018 [30].

Adsorption onto sludge particulates [5], biological treatment [31], chemical treatment [32], photocatalysis [33], electrochemical degradation [34], volatilization [5], and ozonation [35] have been applied for the removal of NP from wastewater. Among the technologies, adsorption has been considered as an important option because of the unique physicochemical properties of NP such as low water solubility and a high log K_{OC} value [3,36].

Recently, carbon nanotubes (CNTs) have received much attention as an adsorbent because of its large surface area, highly porous and hollow structure, light mass density, and strong interaction with hydrophobic organic compounds such as NP [37,38]. On the other hand, spent CNTs can be readily regenerated thermally, chemically, or biologically, so it can be readily reused.

In the present investigation, NP in water samples, which were collected from source waters for 13 Taiwanese water treatment plants (WTPs), was adsorbed to modified multi-walled CNTs (MWCNTs); the kinetics and isotherm equilibrium of the adsorption process were studied. In particular, the influence of different experimental parameters, i.e., water pH, MWCNT dose, contact time, and temperature on the adsorption process.

2. Material and Method

2.1. Materials

Nonyl phenol standard was purchased from Sigma Aldrich (Saint Louis, MS, USA). Multi-walled carbon nanotubes of reagent grade were acquired from Kuang-Yuan Biochemistry Technology (New Taipei City, Taiwan). All other chemicals used in this study are of reagent grade and were also purchased from Sigma Aldrich (Saint Louis, MO, USA). N_2 of 99.99% purity was bought from Shinn Hwa Gas (Taipei, Taiwan). Table 1 summarizes the general properties of NP and MWCNT used in this study. The surface area and pore volume of the MWCNTs were 122 m^2/g and 0.44 cm^3/g, respectively.

Table 1. Characteristics of NP and MWCNTs.

Nonylphenol [39]		CNTs	
Characteristics		Characteristics	
CAS registry number	84852-15-3	Type	MWCNT
Synonyms	p-nonylphenol, 4-nonylphenol,	Length	5–15 μm
Molecular formula	C_9H_{19}-C_6H_5O	Purity	≥95% (v/v)
Molecular weight (g/mol)	220.35	Amorphous carbon	<3%
pK_a	10.7	Ash	≤0.2% (w/w)
log K_{ow}	4.8–5.3	Brunauer–Emmett–Teller (BET) Surface Area	122.2 m^2/g
Henry's law constant (Pa m^3/mol)	11.2	Interlayer distance	0.34 nm
Solubility (mg/L)	5.4–8		

The purchased MWCNTs were pretreated by HNO_3 solution (65%) at 120 °C for 40 min to remove metal impurities from the surface of the MWCNTs. The treated MWCNT samples were, then, washed with deionized (DI) water to remove excess HNO_3, which was carried out by refluxing the solution with MWCNTs in an ultrasonic cleaning bath at 80 °C for 2 h. Ultrasonication treatment also helps to remove amorphous carbons from the MWCNTs. Finally, the treated materials were filtered through a glass fiber filter (GC-50, Advantec, Taipei, Taiwan) to harvest modified MWCNTs.

2.2. Water Sample Preparation

Raw water samples were collected at the source water for each of 13 WTPs in Taiwan (2009–2015). Produced water samples were also collected from the 13 plants. As soon as they were collected, the water samples were filtered with a cellulose acetate membrane filter (0.22–0.45 µm, Merck, Taipei, Taiwan). After filtration, samples were acidified to pH 4.0 using 2 N sulfuric acid and kept at 4 °C.

2.3. Instrumentation and Techniques

The surface morphology of the modified MWCNTs was obtained by a scanning electron microscope (SEM, FEI Quanta 200 Environmental Scanning Electron Microscope, Hillsboro, OR, USA) and a transmission electron microscope (TEM, JEM-1400 Transmission Electron Microscope, Peabody, MA, USA). The zeta potential of the MWCNTs during modification at different pHs (1–11) was measured by a zeta analyzer (Brookhaven BI-90 Plus Particle Size Analyzer, Champaign, IL, USA). Using the BET surface area analyzer (Beckman Coulter SA3100 Surface Area Analyzer, Indianapolis, IN, USA), the surface area and pore size of the modified MWCNTs were analyzed. Nonyl phenol in water samples was extracted by solid phase extraction (SPE) and subsequently quantified by high-performance liquid chromatography-mass spectrometry (HPLC-MS) [40]. The SPE was carried out according to the method proposed by Lin and Tsai [41]. Briefly, the Oasis HLB SPE cartridge (500 mg, 6 mL, Waters, Milford, MA, USA) which was first pretreated with methanol and deionized water of 6 mL each. Then, a water sample of 400 mL with $^{13}C_6$-sulfamethazine (use as a surrogate) added were loaded to the cartridge at 3–6 mL/min. After a sample loaded, the cartridge was rinsed with 6 mL DI water and dried with N_2 gas. Then, analytes were eluted with 4 mL methanol and 4 mL methanol-diethyl ether (1:1 ratio, 50:50, v/v). The eluates were collected to pass for drying with N_2 gas and reconstituted in 0.4 mL of 25% aqueous methanol. Finally, the obtained liquid was filtered through a 0.45 µm PVDF membrane before the HPLC-MS/MS analysis.

2.4. Batch Adsorption Experiments

In the present investigation, all experiments were carried in a batch mode. In the adsorption isotherm experiments, 10 mL of 2.5 mg/L NP was introduced into a centrifuge tube containing different concentrations (0.5–10 mg/L) of adsorbate. The tube was, then, tightly closed with a Teflon-lined screw cap and shaken at 150 rpm at 25 ± 1 °C over 24 h to reach equilibrium. To study the pH dependency of the adsorption, pH (2–10) of the mixture was adjusted by adding 0.1 M HCl/NaOH solution. Once equilibrium was reached, the adsorbent (i.e., MWCNTs) was separated from the solution using a 0.2-µm membrane filter, and the residual NP concentration was analyzed. For the case of kinetic study, samples were collected after 0.5, 1, 2, 4, 7, 8, 10, 12 and 14 h of reaction, and residual NP concentration of each sample was analyzed. All the experiments were performed three times to determine averages and standard deviations; in fact, the calculated c.v. for each set of experiments was within ±3%.

3. Result and Discussion

3.1. NP Concentrations of the Source and Treated Water Samples

As shown in Figure 1, water samples were collected at various locations covering the whole area of Taiwan, including Ban-shin (A), Bao-shan (B), Bei-dou (C), Carp Lake (D), Chan-shin (E), Chen-chin Lake (F), Dong-shin (G), Feng-shan (H), Feng-yuan (I), Kao-tan (J), and Kin-men (K), Nan-hua (L) and Shin-shan (M). All of them are important source waters of Taiwan, so each of them is withdrawn to a nearby WTP. Raw water and produced water samples were collected 2 to 19 times in 2009–2015.

(a)

(b)

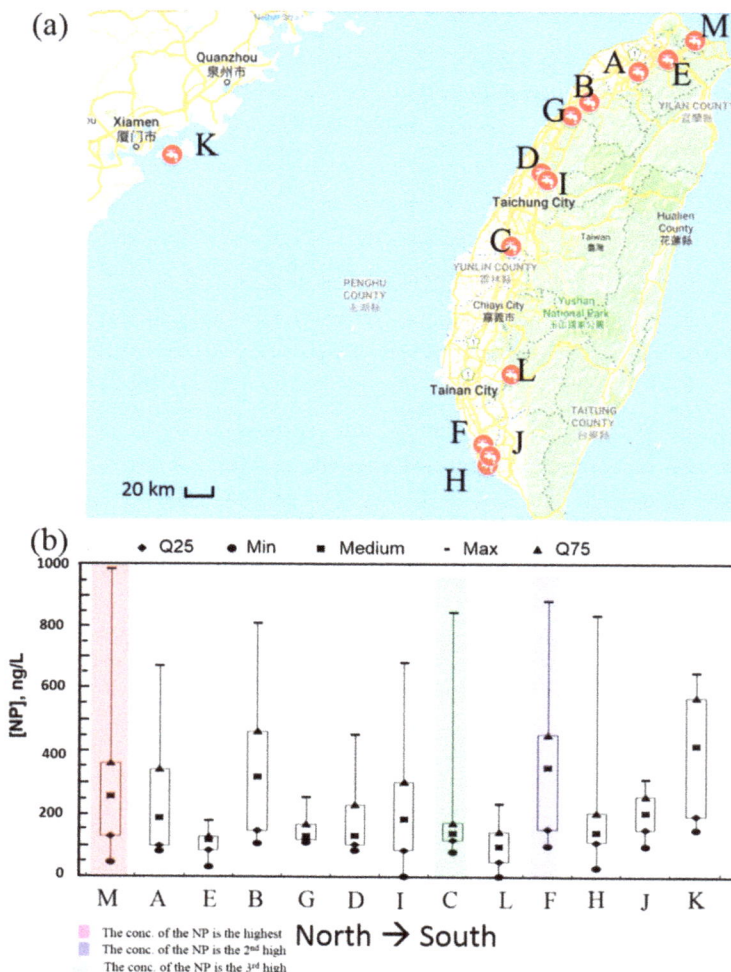

Figure 1. (a) Sampling Locations and (b) NP concentration of source waters. The highest, second highest, and the third highest NP concentrations were detected at (M) Shin-shan, (F) Chen-chin Lake, and (C) Bei-dou, respectively.

The NP concentration of each source water is provided in Supporting Information (Table S2). Different unit processes, namely, rapid filtration, pre-chlorination, coagulation/sedimentation, pre-ozonation, post-chlorination, granular activated carbon, ultrafiltration, and low-pressure reverse osmosis, are employed in combination at each of the WTPs (Table S2).

The result shows that NP concentration was in the range from below detection limit to several hundred ng/L for most of the treatment plants (Figure 1). The maximum NP concentration of 986 ng/L was found at the source water of the Site-M, which consists of conventional processes such as pre-chlorination, coagulation/sedimentation, rapid filtration, and post-chlorination. While the range of NP concentrations of source water was found from 46.1 to 986 ng/L, that of produced water (after conventional treatments) was from below detection limit to 117 ng/L, respectively (as shown in Table S2). It was observed that the NP concentration was not varying significantly at the two sites, i.e., Sites-F and -E. On the other hand, the water samples collected from the Sites-I and -L did not show NP; the concentration was below the detection limit. The high NP contents of the raw water samples collected from Sites-M, -B, -E and -H were attributed to discharge from municipal WWTPs and industrial parks [42].

3.2. Characterization of Adsorbent

The physical properties of MWCNTs, such as specific surface area, pore size distribution and pore volume, were modified via acidification and summarized in Table 2. The acidification increased the surface area, average pore diameter and pore volume of the MWCNTs. Probably, amorphous carbons were removed from the structure by the acidification to open the pores of MWCNTs. While the S_{BET} of the modified MWCNTs was comparable with those reported by others [43–45], the pore volume was much higher; 0.81 cm^3/g was obtained (Table 2). HNO$_3$ treatment reduced metals on the surface of MWCNTs and created new acidic sites, which dominate the roughness of the MWCNTs [46].

Table 2. Specific surface area and pore diameter of MWCNTs.

Adsorbent	S_{BET} (m^2/g)	Pore Diameter (nm)	Pore Volume (cm^3/g)	References
MWCNTs	122	15.1	0.44	
Modified MWCNTs (Acidified)	211	15.5	0.81	
Acidified MWCNTs	237	15.7	0.0042	[44]
Acidified MWCNTs	121		0.49	[45]
Acidified MWCNTs	295	20–40	0.21	[47]
Acidified MWCNTs	254			[48]

Figure 2a gives the SEM (left) and TEM (right) images of raw MWCNTs, respectively. It can be observed that raw MWCNTs shows typical tubular structure morphology, with a length of 5–15 nm. Figure 2b shows the morphology of the modified MWCNTs. The result shows the tubular structure morphology of the modified MWCNTs that has no particle on its surface. The TEM image presents the tubular structure of the raw MWCNTs (right of Figure 2a) and the modified MWCNTs samples (right of Figure 2b). These TEM images confirm the successful removal of metals from the MWCNT surface, which is consistent with the SEM observation.

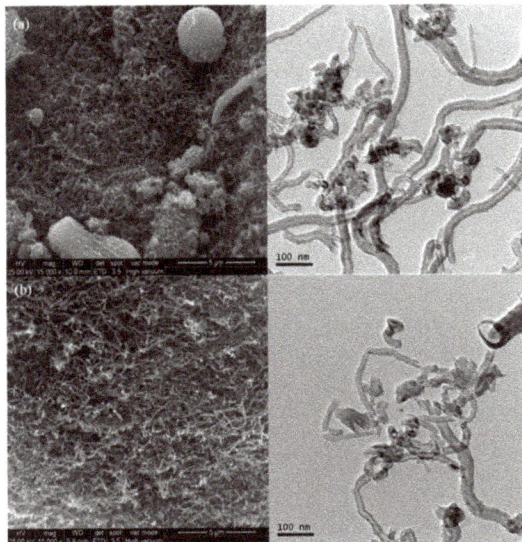

Figure 2. SEM (left) and TEM (right) images of the MWCNTs (**a**) before and (**b**) after modification.

Figure 3 shows the zeta potential of the modified MWCNTs as a function of pH which gives the point of zero-charge (ZPC); the ZPC was observed at pH 2.3 (=pH$_{zpc}$). The zeta potential of the modified MWCNTs was negative in the neutral pH range (Figure 3), indicating that the surface of the MWCNTs was negatively charged. In comparison, the zeta potential of the modified MWCNTs was positive at acidic pHs.

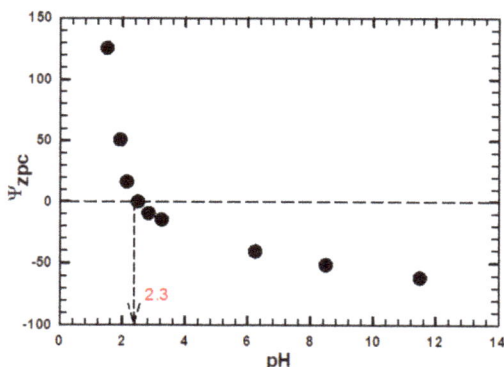

Figure 3. Zeta potential of modified MWCNTs for different pHs.

3.3. Effects of pH and MWCNTs Concentration on NP Removal

Figure 4a clearly shows the influence of pH on the NP removal. At alkaline pHs, NP adsorption decreased. The obtained results are in line with previously obtained ones that deprotonation of organic molecules (phenolic) took place and promoted diffusions [49]. Considering pH$_{zpc}$ of the modified MWCNTs (i.e., 2.3) and pK$_a$ of NP (i.e., 10.7; [39]), at neutral pHs (e.g., 4–10), the adsorption of NP onto the MWCNTs is favored due to the weak electrostatic interaction and H-bond formation between the negatively charged MWCNTs and NP. Figure 4b demonstrates the effect of the modified MWCNTs dosage (0.1 to 10 mg/L) on NP adsorption at pH 4. As the MWCNTs concentration increased, the NP adsorption increased to some extent possibly due to increased free binding sites. As shown in Figure 4b, the highest adsorption capacity of the modified MWCNTs for NP was estimated at 1040 mg/g at pH 4 and 25 °C. In any case, the adsorbent mass higher than 0.5 mg/L did not further improve the adsorption. Therefore, the modified MWCNTs of 0.5 mg/L was used in the following batch experiments.

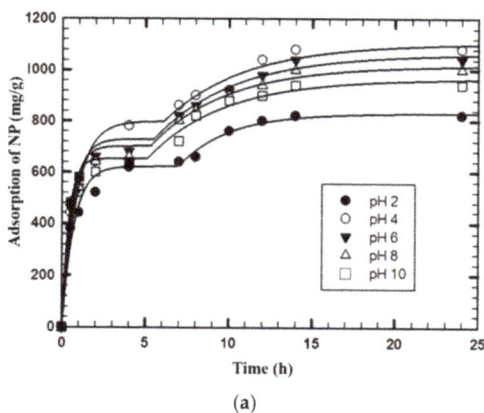

(a)

Figure 4. *Cont.*

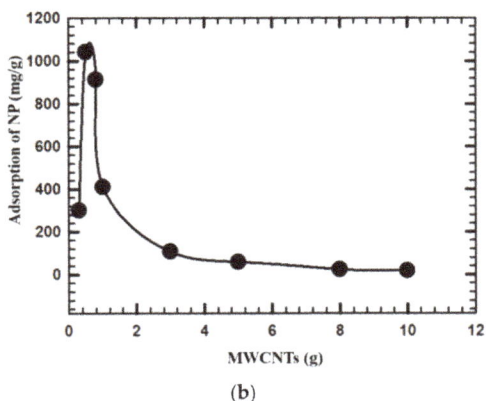

Figure 4. Effect of (**a**) pH ([MWCNT] = 0.5 mg/L) and (**b**) adsorbent (MWCNTs) dose (at pH 4) on the adsorption of NP by MWCNTs. Experimental conditions: [NP] = 2.5 mg/L, T = 25 °C).

3.4. Adsorption Kinetics

Figure 5 shows the kinetics of NP adsorption onto the modified MWCNTs. The obtained result shows that the equilibrium is established in 14 h (Figure 4a). In the beginning, the rate of adsorption was high probably due to a large number of free surface sites available on the MWCNTs. Afterward, the adsorption rate became slower and finally reached the equilibrium state. Thus, it could be concluded that free binding sites and affinity between the modified MWCNT and NP might be responsible for the NP adsorption, similar to the result from our previous study in which NP was adsorbed to clays and soil materials [50]. In fact, the kinetics profile from the current NP adsorption study is characterized with two linear regions; they are the liquid film diffusion layer and the intra-particle diffusion one [51]. The rate of NP adsorption in the liquid film diffusion layer is expressed as Equation (1):

$$\frac{dq}{dt} = k_f S_0 (C - C_i) \tag{1}$$

where S_0 is the particle surface area per unit particle volume, and k_f is the rate constant.

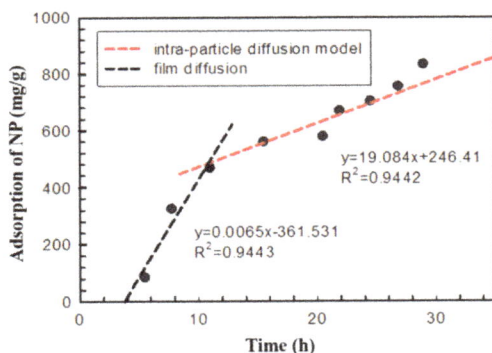

Figure 5. The kinetics of NP adsorption described by the film diffusion and the intra-particle diffusion models. Experimental conditions: [NP] = 2.5 mg/L, T = 25 °C, pH = 4, and [MWCNT] = 0.5 mg/L.

The intra-particle diffusion model assumes that the adsorption density, q_t, varies proportionally with $t^{\frac{1}{2}}$, as shown in Equation (2),

$$q_t = k_{id}\sqrt{t} + I \tag{2}$$

where I and k_{id} is the intercept and the intra-particle diffusion rate constant, respectively. The boundary layer effect is indicated by the initial high slope, and the intra-particle diffusion or pore diffusion is demonstrated by the less steep slope (Figure 5).

The NP adsorption capacity increases with the increase in the adsorbent surface area (S_0) [45]. Moreover, the increase in the surface area of the adsorbent strongly correlates with the equilibrium adsorption capacity [49]. This kind of relationship has already been described in the literature [52]. It suggests that the pore-filling mechanism is mostly responsible for the adsorption. Generally, small-size organic compounds are responsible for pore-filling [43,53]. In addition, the octanol-water partitioning coefficient (log K_{ow}), an indicator of hydrophobicity of an organic compound, plays an important role in the adsorption [1]. Therefore, the compound with a high log K_{ow} (i.e., log K_{ow} > 4) should exhibit a strong adsorption affinity towards the active binding sites of the adsorbent [1]. Since the log K_{ow} of NP is 4.8–5.3 (Table 1), NP seems to have a strong adsorption affinity.

Table 3 lists the kinetics models that were applied for the NP adsorption to the modified MWCNTs. The coefficients of determination for the kinetics models are in the order of Elovich (R^2 = 0.96) > pseudo-first order (0.93) > pseudo-second order (0.46).

Table 3. Kinetic models and parameters for adsorption of NP to MWCNTs.

Kinetic model		Parameter	Value	R^2
Pseudo-first-order	$log(q_e - q_t) = log q_e - \frac{k_1}{2.303}t$	q_e (mg/g)	2.1×10^1	0.93
		k_1 (/min)	2.7×10^{-1}	
Pseudo-second-order	$\frac{t}{q_t} = \frac{1}{k_2 q_e^2} + \frac{1}{q_e}t$	q_e (mg/g)	3×10^2	0.46
		k_2 (g/mg/min)	1.1×10^{-3}	
Elovich	$q = \left(\frac{2.3}{\alpha}\right)log(t + t_0) - \left(\frac{2.3}{\alpha}\right)\log t_0$ $t_0 = \frac{1}{a\alpha}$	a	4.3×10^{-3}	0.96
		α (g/mg/min)	4.3×10^2	
Double exponential model (DEM)	$q_t = q_e - \frac{D1}{m_a}exp^{(-Dk_1 t)}$ $- \frac{D2}{m_a}exp^{(-Dk_2 t)}$	D_1 (mg/L)	8.4×10^{-3}	–
		Dk_1 (/s)	1.4×10^{-3}	
		D_2 (mg/L)	8.5×10^{-3}	
		Dk_2 (/s)	1.4×10^{-3}	

Note: q_e and q_t are the amount of NP adsorbed at equilibrium at time t (mg/g), respectively. k_1 and k_2 are the rate constant of the pseudo-first-order, pseudo-second order, respectively. a is the Elovich constant and α is the initial sorption rate constant for the Elovich model (g/mg/min). D_1 and D_2 are the 1st and 2nd stage concentrations (mg/L) in the DEM model, while Dk_1, and Dk_2 are the 1st and 2nd stage rate constants (1/s). Lastly, m_a is dosage of an adsorbent in the DEM model (g/L).

The calculated q_e value was not in agreement with the experimental data, indicating that the pseudo-second-order kinetic model may not explain the adsorption of NP onto the modified MWCNTs. Preferably, the experimental adsorption capacity data well fit to the Elovich model, which implies that the NP adsorption is governed by chemisorption processes [54].

3.5. Equilibrium Adsorption Isotherm

To examine the effect of the initial NP concentration on the adsorption capacity of the modified MWCNTs, the adsorption isotherm study was carried out. Table 4 shows the adsorption isotherm models applied for the adsorption of NP onto MWCNTs. The Langmuir [55], Freundlich [56], Jovanovic [57], Elovich [54] and Temkin [58] models were respectively employed to fit the adsorption data, and the result is summarized (Table 4).

Table 4. Parameters of NP adsorption based on various models.

Isotherm model		Parameter	Value	R^2	Reference
Langmuir	$q_e = \frac{k_L \times q_m \times C_e}{1 + k_L C_e}$	q_m (mg/g)	1.11×10^4	0.9763	[55]
		K_L (L/mg)	1.5		
Freundlich	$q_e = K_f C_e^{\frac{1}{n_f}}$	K_f ((mg/g) (mg/L)$^{1/n}$)	33.20	0.994	[56]
		n_f	0.89		
Temkin	$q_e = BlnA_T + BlnC_e$	C_e (g/mg)	6.03×10^4	0.9986	[57,59]
		A_T (L/mg)	0.60		
Elovich	$q_e = K_E q_m C_e e^{\frac{q_e}{q_m}}$	q_m (mg/g)	2.5×10^4	0.9996	[53]
		K_E (L/mg)	0.38		
Jovanovic	$q_e = q_m \left(1 - e^{K_i C_e}\right)$	K_j (L/mg)	0.22	0.9984	[57]
		q_m (g/mg)	47.37		

In short, Freundlich, Jovanovic, Temkin, and Elovich models fitted the data better than the Langmuir. The maximum NP adsorption capacity (q_{max}, mg/g) of the modified MWCNTs estimated using the Langmuir model at pH 2 and 293 K was 1.11×10^4 mg/g. On the other hand, a much higher q_{max} value (i.e., 25,000 mg/g) was estimated by the Elovich isotherm. Furthermore, the calculated adsorption data matched well with the observed ones. The current result demonstrated that the adsorption of NP onto the modified MWCNTs is multilayer (chemisorption) rather than Langmuir-mono-layer.

3.6. Thermodynamic Aspects

Temperature affects the adsorption capacity of an adsorbent. It was observed that the amount of NP adsorbed increased as temperature increased, indicating an endothermic process. To evaluate the spontaneity of the adsorption process, the thermodynamic parameters, i.e., changes in the Gibb's free energy ($\Delta G°$, kJ/mol), the enthalpy ($\Delta H°$, kJ/mol), and the entropy ($\Delta S°$, kJ/mol/K) of the process were calculated by Equations (3)–(5):

$$\Delta G = -R\,T\,\ln K_L \tag{3}$$

$$\Delta G° = \Delta H° - T\,\Delta S°, \tag{4}$$

$$\ln K_L = \frac{\Delta S°}{R} - \frac{\Delta H°}{RT} \tag{5}$$

where R is the universal gas constant (8.314×10^{-3} kJ/mol·K), and T is the absolute temperature (K). The K_L (L/g), equilibrium constant, is calculated from the Langmuir constant, q_m [55]. The value of $\Delta G°$ was calculated from Equation (3), while $\Delta H°$ and $\Delta S°$ were estimated from the slop and intercept of the graph (straight line) of ((1/T) vs. ln K) (Figure S2). The values of various thermodynamic parameters for the NP adsorption onto the modified MWCNTs are presented in Table 5.

Table 5. Values of various thermodynamic parameters for adsorption of NP to MWCNTs.

Temp. (°C)	K (L/mol)	$\Delta G°$ (kJ/mol)	$\Delta H°$ (kJ/mol)	$\Delta S°$ (J/K mol)
15	1.47	−0.93	11.18	42.02
25	1.72	−1.35		
35	2.00	−1.77		
45	2.29	−2.19		

The negative $\Delta G°$ values suggest that the adsorption of NP onto the MWCNTs is spontaneous. The $\Delta G°$ value negatively increases as the temperature increases. This indicates that the adsorption is more favorable at a higher temperature from the thermodynamics point of view (values of K_L increased from 1.47 to 2.29). Additionally, the $\Delta H°$ and $\Delta S°$ values are all positive, indicating the NP adsorption onto the modified MWCNTs is endothermic at the temperature range of 288–318 K.

From the adsorption and kinetics data, the following three important factors governing the adsorption of NP onto the modified MWCNTs can be drawn; they are (1) physical properties of the adsorbent, i.e., size, shape, surface area, impurities, functional group, and pore size; (2) environmental condition, i.e., pH; (3) π–π interaction between the MWCNTs and NP. In this study, the physical properties of MWCNTs were improved by acidification, which renders the modified MWCNTs surface acidic. In addition to the physical properties, relatively high pK_a and K_{ow} values (hydrophobicity) of NP might enhance the interaction between ions/compounds with the modified MWCNTs. Finally, the hydrogen-bond formation was observed between -H of the carbonyl group on the modified MWCNTs and the –OH group of NP. Therefore, H-bond might be a key mechanism responsible for the NP adsorption in this study. In particular, the H and O contents were higher in the modified MWCNTs compared with pristine MWCNTs.

4. Conclusions

In this study, the occurrence of NP in 13 sites of Taiwan from 2009 to 2015 has been comprehensively investigated. The concentration of NP was habitually distinguished at a critical level in Taiwan surface water, especially in 12 of 13 investigated source waters, indicating the traditional water treatment processes are not effective.

Using modified (acid treated) MWCNTs as an adsorbent may be advantageous in future for removal of NP due to its higher surface area and specific surface functionality. The highest NP adsorption capacity on MWCNTs was 1040 mg/g at initial NP concentration of 2.5 mg/L at pH 4. The adsorption process was found to follow the Elovich kinetics model and the Elovich isotherm model, implying that the NP adsorption onto MWCNTs is a multilayer. Three factors namely, physical properties of the adsorbent, environmental conditions, and interaction of the adsorbent with NP are considered as the governing factors for the adsorption process. Also, a result of the thermodynamic analysis shows that the adsorption is endothermic and depends on the solution temperature. In short, all the obtained results indicate that the acidified MWCNTs can be employed successfully for the removal of NP from aqueous solution.

Supplementary Materials: The following are available online at http://www.mdpi.com/2076-3417/8/11/2295/s1, Figure S1: Adsorption kinetics equation of NP. Experimental conditions: [NP] = 2.5 mg/L, T = 25 °C, [MWCNTs] = 0.5 mg/L, Figure S2: Plot of ln (K) versus 1/T (van't Hoff plot) for various temperature from 288 K to 318 K, [NP] = 2.5 mg/L and [MWCNTs] = 0.5 mg/L, Table S1: The concentration level of NP in various Rivers of Asian Region; Table S2: The concentration of Nonylphenol in Public Water Supply Systems (from 2009 to 2015).

Author Contributions: All the authors equally contributed to this manuscript.

Funding: The authors wish to thank the Ministry of Science and Technology (MOST) of Taiwan (R.O.C.) for the financial support under grant number MOST 107-3113-E-007-002. H. Kim is is financially supported by Korea Environmental Industry & Technology Institute (2015001790002).

Conflicts of Interest: The authors declare no conflict of interest.

References

1. Soares, A.; Guieysse, B.; Jefferson, B.; Cartmell, E.; Lester, J.N. Nonylphenol in the environment: A critical review on oCcurrence, fate, toxicity and treatment in wastewaters. *Environ. Int.* **2008**, *34*, 1033–1049. [CrossRef] [PubMed]
2. Ferrara, F.; Ademollo, N.; Delise, M.; Fabietti, F.; Funari, E. Alkylphenols and their ethoxylates in seafood from the Tyrrhenian sea. *Chemosphere* **2008**, *72*, 1279–1285. [CrossRef] [PubMed]

3. Groshart, C.P.; Okkerman, P.C.; Wassenberg, W.B.A.; Pijnenburg, A.M.C.M. Chemical study on alkylphenols. In *RIKZ Report 2001.029*; Nat. Inst. Coast and Marine Management (RIKZ): The Hague, The Netherlands, 2001.

4. Careghini, A.; Mastorgio, A.F.; Saponaro, S.; Sezenna, E. Bisphenol a, nonylphenols, benzophenones, and benzotriazoles in soils, groundwater, surface water, sediments, and food: A review. *Environ. Sci. Pollut. Res.* **2015**, *22*, 5711–5741. [CrossRef] [PubMed]

5. Bouki, C.; Dvorakova, M.; Diamadopoulos, E. Adsorption of nonylphenol on activated sludge biomass under aseptic conditions. *Clean Soil Air Water* **2010**, *38*, 516–520. [CrossRef]

6. Mao, Z.; Zheng, X.-F.; Zhang, Y.-Q.; Tao, X.-X.; Li, Y.; Wang, W. Occurrence and biodegradation of nonylphenol in the environment. *Int. J. Mol. Sci.* **2012**, *13*, 491–505. [CrossRef] [PubMed]

7. Duong, C.N.; Ra, J.S.; Cho, J.; Kim, S.D.; Choi, H.K.; Park, J.H.; Kim, K.W.; Inam, E.; Kim, S.D. Estrogenic chemicals and estrogenicity in river waters of south Korea and seven Asian countries. *Chemosphere* **2010**, *78*, 286–293. [CrossRef] [PubMed]

8. Chen, H.W.; Liang, C.H.; Wu, Z.M.; Chang, E.E.; Lin, T.F.; Chiang, P.C.; Wang, G.S. Occurrence, and assessment of treatment efficiency of nonylphenol, octylphenol and bisphenol-A in drinking water in Taiwan. *Sci. Total Environ.* **2013**, *449*, 20–28. [CrossRef] [PubMed]

9. Derbalah, A.S.H.; Nakatani, N.; Sakugawa, H. Distribution, seasonal pattern, flux and contamination source of pesticides and nonylphenol residues in Kurose River water, Higashi-Hiroshima, Japan. *GeoChem. J.* **2003**, *37*, 217–232. [CrossRef]

10. Li, D.; Kim, M.; Oh, J.R.; Park, J. Distribution characteristics of nonylphenols in the artificial Lake Shihwa, and surrounding creeks in Korea. *Chemosphere* **2004**, *56*, 783–790. [CrossRef] [PubMed]

11. Li, D.; Kim, M.; Shim, W.J.; Yim, U.H.; Oh, J.R.; Kwon, Y.J. Seasonal flux of nonylphenol in Han River, Korea. *Chemosphere* **2004**, *56*, 1–6. [CrossRef] [PubMed]

12. Kawahata, H.; Ohta, H.; Inoue, M.; Suzuki, A. Endocrine disrupter nonylphenol and bisphenol A contamination in Okinawa and Ishigaki Islands, Japan—Within coral reefs and adjacent river mouths. *Chemosphere* **2004**, *55*, 1519–1527. [CrossRef] [PubMed]

13. Cheng, C.Y.; Wu, C.Y.; Wang, C.H.; Ding, W.H. Determination and distribution characteristics of degradation products of nonylphenol polyethoxylates in the rivers of Taiwan. *Chemosphere* **2006**, *65*, 2275–2281. [CrossRef] [PubMed]

14. Xu, J.; Wang, P.; Guo, W.; Dong, J.; Wang, L.; Dai, S. Seasonal and spatial distribution of nonylphenol in Lanzhou Reach of Yellow River in China. *Chemosphere* **2006**, *65*, 1445–1451. [CrossRef] [PubMed]

15. Fu, M.; Li, Z.; Gao, H. Distribution characteristics of nonylphenol in Jiaozhou Bay of Qingdao and its adjacent rivers. *Chemosphere* **2007**, *69*, 1009–1016. [CrossRef] [PubMed]

16. Hong, L.; Li, M.H. Acute Toxicity of 4-Nonylphenol to Aquatic Invertebrates in Taiwan. *Bull. Environ. Contam. Toxicol.* **2007**, *78*, 445–449. [CrossRef] [PubMed]

17. Motegi, M.; Nojiri, K.; Hosono, S.; Kawamura, K. Determination and evaluation of estrogenic contamination in an urban river basin. *J. Environ. Chem.* **2007**, *17*, 421–431. (In Japanese with English Abstract). [CrossRef]

18. Wu, Z.; Zhang, Z.; Chen, S.; He, F.; Fu, G.; Liang, W. Nonylphenol and octylphenol in urban eutrophic lakes of the subtropical China. *Fresenius Environ. Bull.* **2007**, *16*, 227–234.

19. Hong, S.; Won, E.J.; Ju, H.J.; Kim, M.S.; Shin, K.H. Current nonylphenol pollution and the past 30 years record in an artificial Lake Shihwa, Korea. *Mar. Pollut. Bull.* **2010**, *60*, 308–313. [CrossRef] [PubMed]

20. Shue, M.F.; Chen, F.A.; Chen, T.C. Total estrogenic activity and nonylphenol concentration in the Donggang River. *Taiwan. Environ. Monit. Assess.* **2010**, *168*, 91–101. [CrossRef] [PubMed]

21. Cherniaev, A.P.; Kondakove, A.S.; Zyk, E.N. Contents of 4-Nonylphenol in Surface Sea Water of Amur Bay (Japan/East Sea). *Achiev. Life Sci.* **2016**, *10*, 65–71. [CrossRef]

22. Diao, P.; Chen, Q.; Wang, R.; Sun, D.; Cai, Z.; Wu, H.; Duan, S. Phenolic endocrine-disruptiong compounds in the Pearl River Estuary: Occurrence, bioaccumulation and risk assessment. *Sci. Total Environ.* **2017**, *584–585*, 1100–1107. [CrossRef] [PubMed]

23. Watanabe, M.; Takano, T.; Nakamura, K.; Watanabe, S.; Seino, K. Water quality and concentration of alkylphenols in river used as source of drinking water and flowing through urban areas. *Environ. Health Prev. Med.* **2017**, *12*, 17–24. [CrossRef] [PubMed]

24. Wu, M.; Wang, L.; Xu, G.; Liu, N.; Tang, L.; Zheng, J.; Bu, T.; Lei, B. Seasonal and spatial distribution of 4-tert-oCtylphenol, 4-nonylphenol and bisphenol A in the Huangpu River and its tributaries, Shanghai, China. *Environ. Monit. Assess.* **2013**, *185*, 3149–3161. [CrossRef] [PubMed]

25. Xu, E.G.B.; Liu, S.; Ying, G.G.; Zheng, G.J.S.; Lee, J.H.W.; Leung, K.M.Y. The occurrence and ecological risks of endocrine disrupting chemicals in sewage effluents from three different sewage treatment plants, and in natural seawater from a marine reserve of Hong Kong. *Mar. Pollut. Bull.* **2014**, *85*, 352–362. [CrossRef] [PubMed]

26. Zhang, Z.F.; Ren, N.Q.; Kannan, K.; Nan, J.; Liu, L.Y.; Ma, W.L.; Qi, H.; Li, Y.F. Occurrence of Endocrine-Disrupting Phenols and Estrogens in Water and Sediment of the Songhua River, Northeastern China. *Arch. Environ. Contam. Toxicol.* **2014**, *66*, 361–369. [CrossRef] [PubMed]

27. Zhong, M.; Yin, P.; Zhao, L. Nonylphenol and octylphenol in riverine waters and surface sediments of the Pearl River Estuaries, South China: Occurrence, ecological and human health risks. *Water Sci. Technol. Water Supply* **2017**, *17*, 1070–1079. [CrossRef]

28. Cheng, J.R.; Wang, K.; Yu, J.; Yu, Z.X.; Yu, X.B.; Zhang, Z.Z. Distribution and fate modeling of 4-nonylphenol, 4-t-oCtylphenol, and bisphenol A in the Yong River of China. *Chemosphere* **2018**, *195*, 594–605. [CrossRef] [PubMed]

29. Zhang, Y.Z.; Tang, C.Y.; Song, X.F.; Li, F.D. Behavior and fate of alkylphenols in surface water of the Jialu river, Henan province, china. *Chemosphere* **2009**, *77*, 559–565. [CrossRef] [PubMed]

30. CPAMI, Construction and Planning Agency Ministry of Interior. 2018. Available online: https://www.cpami.gov.tw/public-information/laws-regulations.html (accessed on 20 July 2018).

31. Joss, A.; Andersen, H.; Ternes, T.; Richle, P.R.; Siegrist, H. Removal of estrogens in municipal wastewater treatment under aerobic and anaerobic conditions: Consequences for plant optimization. *Environ. Sci. Technol.* **2004**, *38*, 3047–3055. [CrossRef] [PubMed]

32. Tsutsumi, Y.; Haneda, T.; Nishida, T. Removal of estrogenic activities of bisphenol a and nonylphenol by oxidative enzymes from lignin-degrading basidiomycetes. *Chemosphere* **2001**, *42*, 271–276. [CrossRef]

33. Dzinun, H.; Othman, M.H.D.; Ismail, A.F.; Puteh, M.H.; Rahman, M.A.; Jaafar, J. PhotoCatalytic degradation of nonylphenol by immobilized tio2 in dual layer hollow fibre membranes. *Chem. Eng. J.* **2015**, *269*, 255–261. [CrossRef]

34. Ciorba, G.A.; Radovan, C.; Vlaicu, I.; Masu, S. Removal of nonylphenol ethoxylates by electrochemically-generated coagulants. *J. Appl. Electrochem.* **2002**, *32*, 561–567. [CrossRef]

35. Barrera-Díaz, C.E.; Frontana-Uribe, B.A.; Rodríguez-Peña, M.; Gomez-Palma, J.C.; Bilyeu, B. Integrated advanced oxidation process, ozonation-electrodegradation treatments, for nonylphenol removal in batch and continuous reactor. *Catal. Today* **2018**, *305*, 108–116. [CrossRef]

36. Rashed, M.N. Adsorption technique for the removal of organic pollutants from water and wastewater. In *Organic Pollutants-Monitoring, Risk and Treatment*; InTech: London, UK, 2013; pp. 167–194.

37. Ren, X.; Chen, C.; Nagatsu, M.; Wang, X. Carbon nanotubes as adsorbents in environmental pollution management: A review. *Chem. Eng. J.* **2011**, *170*, 395–410. [CrossRef]

38. Amin, M.T.; Alazba, A.A.; Manzoor, U. A review of removal of pollutants from water/wastewater using different types of nanomaterials. *Adv. Mater. Sci. Eng.* **2014**, *2014*, 1–24. [CrossRef]

39. Ahel, M.; Giger, W. Aqueous solubility of alkylphenols and alkylphenol polyethoxylates. *Chemosphere* **1993**, *26*, 1461–1470. [CrossRef]

40. Chokwe, T.B.; Okonkwo, J.O.; Sibali, L.L. Distribution, exposure pathways, sources and toxicity of nonylphenol and nonylphenol ethoxylates in the environment. *Water SA* **2017**, *42*, 529–542. [CrossRef]

41. Lin, A.Y.-C.; Tsai, Y.-T. Occurrence of pharmaceuticals in taiwan's surface waters: Impact of waste streams from hospitals and pharmaceutical production facilities. *Sci. Total Environ.* **2009**, *407*, 3793–3802. [CrossRef] [PubMed]

42. Aerni, H.R.; Kobler, B.; Rutishauser, B.V.; Wettstein, F.E.; Fischer, R.; Giger, W.; Hungerbuhler, A.; Marazuela, M.D.; Peter, A.; Schonenberger, R.; et al. Combined biological and chemical assessment of estrogenic activities in wastewater treatment plant effluents. *Anal. Bioanal. Chem.* **2004**, *378*, 688–696. [CrossRef] [PubMed]

43. Wang, S.G.; Liu, X.W.; Gong, W.X.; Nie, W.; Gao, B.Y.; Yue, Q.Y. Adsorption of fulvic acids from aqueous solutions by carbon nanotubes. *J. Chem. Technol. Biotechnol.* **2007**, *82*, 698–704. [CrossRef]

44. Wang, H.J.; Zhou, F.P.; Yu, H.; Chen, L.F. Adsorption characteristic of acidified carbon nanotubes for heavy metal Pb(II) in aqueous solution. *Mater. Sci. Eng. A* **2007**, *466*, 201–206. [CrossRef]

45. Liao, Q.; Sun, J.; Gao, L. The adsorption of resorcinol from water using multi-walled carbon nanotubes. *Colloids Surf. A* **2008**, *312*, 160–165. [CrossRef]

46. Reinert, L.; Lasserre, F.; Gachot, C.; Grutzmacher, P.; MacLucas, T.; Souza, M.; Mucklich, F.; Suarez, S. Long-lasting solid lubrication by CNT-coated patterned surfaces. *Sci. Rep.* **2017**, *42873*, 1–13. [CrossRef] [PubMed]

47. Lu, C.; Chung, Y.-L.; Chang, K.-F. Adsorption of trihalomethanes from water with carbon nanotubes. *Water Res.* **2005**, *39*, 1183–1189. [CrossRef] [PubMed]

48. Wang, H.; Zhou, A.; Peng, F.; Yu, H.; Yang, J. Mechanism study on adsorption of acidified multiwalled carbon nanotubes to Pb(II). *J. Colloide Interface Sci.* **2007**, *316*, 277–283. [CrossRef] [PubMed]

49. Shah, K.J.; Pan, S.-Y.; Shukla, A.D.; Shah, D.O.; Chiang, P.C. Mechanism of organic pollutant sorption from aqueous solution by cationic tunable organoClays. *J. Colloid Interface Sci.* **2018**, *529*, 90–99. [CrossRef] [PubMed]

50. Chen, G.; Shah, K.J.; Shi, L.; Chiang, P.C. Removal of cd(ii) and pb(ii) ions from aqueous solutions by synthetic mineral adsorbent: Performance and mechanisms. *Appl. Surf. Sci.* **2017**, *409*, 296–305. [CrossRef]

51. Weber, W.J.; Morris, J.C. Kinetics of adsorption on carbon from solutions. *J. Sanit. Eng. Div.* **1963**, *89*, 30.

52. RoCher, V.; Paffoni, C.; Gonçalves, A.; Guérin, S.; Azimi, S.; Gasperi, J.; Moilleron, R.; Pauss, A. Municipal wastewater treatment by biofiltration: Comparisons of various treatment layouts. Part 1: Assessment of carbon and nitrogen removal. *Water Sci. Technol.* **2012**, *65*, 1705. [CrossRef] [PubMed]

53. Ersan, G.; Kaya, Y.; Apul, O.G.; Karanfil, T. Adsorption of organic contaminants by graphene nanosheets, carbon nanotubes and granular activated carbons under natural organic matter preloading conditions. *Sci. Total Environ.* **2016**, *565*, 811–817. [CrossRef] [PubMed]

54. Elovich, S.Y.; Larionov, O.G. Theory of adsorption from nonelectrolyte solutions on solid adsorbents. *Bull. Acad. Sci. USSR Div. Chem. Sci.* **1962**, *11*, 198–203. [CrossRef]

55. Langmuir, I. The constitution and fundamental properties of solids and liquids part i solids. *J. Am. Chem. Soc.* **1916**, *38*, 2221–2295. [CrossRef]

56. Freundlich, H. Oberflächeneinflüsse beim bler und bei der bierbereitung. *Zeitschrift für Chemie und Industrie der Kolloide* **1906**, *1*, 152. [CrossRef]

57. Jovanović, D.S. Physical adsorption of gases. *Kolloid-Zeitschrift Zeitschrift für Polymere* **1969**, *235*, 1214–1225. [CrossRef]

58. Isik, M. Biosorption of ni(ii) from aqueous solutions by living and non-living ureolytic mixed culture. *Colloids Surf. B Biointerfaces* **2008**, *62*, 97–104. [CrossRef] [PubMed]

59. Basha, S.; Murthy, Z.V.P.; Jha, B. Sorption of hg(ii) from aqueous solutions ontoCarica papaya: Application of isotherms. *Ind. Eng. Chem. Res.* **2008**, *47*, 980–986. [CrossRef]

applied sciences

MDPI

Article

Novel Zeolitic Imidazolate Frameworks Based on Magnetic Multiwalled Carbon Nanotubes for Magnetic Solid-Phase Extraction of Organochlorine Pesticides from Agricultural Irrigation Water Samples

Xiaodong Huang [1], Guangyang Liu [2], Donghui Xu [2], Xiaomin Xu [2], Lingyun Li [2], Shuning Zheng [2], Huan Lin [2] and Haixiang Gao [1,*

[1] Department of Applied Chemistry, China Agricultural University, Beijing 100193, China; huangxiaodong@caas.cn
[2] Institute of Vegetables and Flowers, Chinese Academy of Agricultural Sciences, Key Laboratory of Vegetables Quality and Safety Control, Ministry of Agriculture and Rural Affairs of China, Beijing 100081, China; liuguangyang@caas.cn (G.L.); xudonghui@caas.cn (D.X.); hsuixiaomin@gmai.com (X.X.); lilingyun@caas.cn (L.L.); zhengshuning@caas.cn (S.Z.); linhuan03@caas.cn (H.L.)
* Correspondence: hxgao@cau.edu.cn; Tel.: +86-10-62731991

Received: 18 May 2018; Accepted: 5 June 2018; Published: 12 June 2018

Abstract: Magnetic solid-phase extraction is an effective and convenient sample pretreatment technique that has received considerable interest in recent years. A lot of research indicated that magnetic nanocarbon-material-based composites have good application prospects as adsorbents for magnetic solid-phase extraction of pesticides. Herein, a novel zeolitic imidazolate framework based on magnetic multiwalled carbon nanotubes (M-M-ZIF-67) has been prepared as an adsorbent for magnetic solid-phase extraction of nine organochlorine pesticides from agricultural irrigation water samples. The obtained M-M-ZIF-67 material possessed porous surfaces and super-paramagnetism due to the utilization of magnetic multiwalled carbon nanotubes as the magnetic kernel and support. To evaluate the extraction performance of the M-M-ZIF-67, the main parameters that affected the extraction efficiency were researched. Under the optimal conditions, a good linearity for the nine organochlorine pesticides was achieved with the determination coefficients (R^2) higher than 0.9916. The limits of detection (signal/noise = 3:1) were in the range 0.07–1.03 $\mu g\ L^{-1}$. The recoveries of all analytes for the method at spiked levels of 10 and 100 $\mu g\ L^{-1}$ were 74.9–116.3% and 75.1–112.7%, respectively. The developed M-M-ZIF-67 based magnetic solid-phase extraction method has a potential application prospect for the monitoring of trace level of organochlorine pesticides in environmental water samples.

Keywords: zeolitic imidazolate framework; multi-walled carbon nanotubes; magnetic solid phase extraction; organochlorine pesticides; agricultural irrigation water

1. Introduction

Sample pretreatment is a crucial step in analysis of trace or ultra-trace amounts analytes in complex matrices. Solid-phase extraction (SPE) is a type of widely used pretreatment for effective concentration of analytes in complex matrices before instrumental analysis [1]. A variety of pretreatment methods has been developed based on this technique, including solid-phase microextraction (SPME) [2], micro-SPE (μ-SPE) [3], and stir-bar sorptive extraction (SBSE) [4].

Magnetic solid-phase extraction (MSPE), as a new type of SPE, is a pretreatment method that has received considerable interest in recent years. In this technique, magnetic adsorbents are directly dispersed into sample solutions, and this dispersive extraction mode can enhance the contact area between adsorbents and analytes [5]. Notably, magnetic adsorbents can be separated from the sample solutions under an external magnetic field without the need of traditional centrifugation or filtration, thereby simplifying the extraction process. Furthermore, magnetic adsorbents can be recycled and reused easily, which is cost effective and environmentally friendly [6]. Therefore, the MSPE technique shows comprehensive advantages of simplicity, time, reagent and labor savings, and excellent extraction efficiency, which meet the principles of green analytical chemistry [7]. The diversity of the materials used in MSPE is the main factor that has led to extensive development of this technique in recent decades [8]. Multiple magnetic sorbents have been synthesized by embedding magnetic cores in different organic or inorganic materials, such as chitosan [9], ionic liquids [10], polymers [11], silica [12], metallic oxides [13], molecularly imprinted polymers [14], and carbon nanomaterials [15,16].

Multiwalled carbon nanotubes (MWCNTs) are formed by seamless rolling up of several layers of graphite sheets. Because of their excellent properties, such as high surface area and inner volume, stability, mechanical strength, ability to establish π–π interactions, and capacity for functionalization, MWCNTs have the possibility of acting as good sorbents [17]. MWCNTs have recently attracted considerable interest as adsorbents in MSPE for extracting different analytes, such as antibiotics [18], estrogens [19], mycotoxins [20], metal ions [21], environmental pollutants [22], and pesticides [23,24]. The use of magnetic MWCNTs combined with other materials has attracted great interest.

Metal–organic frameworks (MOFs) are microporous inorganic–organic crystalline structure materials. They are formed by self-assembly of metal ions (clusters or secondary building units) and organic ligands (linkers) by coordination bonds, and they have a highly ordered and three-dimensional structure [25]. MOFs are promising sorbent materials, and using them for extraction could have the advantages of enhanced selectivity and stability, permeable channels and coordination nanospace, framework flexibility and dynamics, easy tunability, and modification [26]. However, application of MOFs is limited in certain cases owing to the lack of water and thermal stability, for example, MOF-199 and MOF-5 lose their extraction efficiency when they are exposed to moisture for a long time [27].

Zeolitic imidazolate frameworks (ZIFs) are a subclass of MOFs that are composed of tetrahedral transition metal ions (e.g., Zn and Co) and imidazolate-type organic linkers, and they exhibit high water stability in aqueous media [28]. Owing to their features of microporosity, uniform structured cavities, and a high surface area, ZIFs have many applications, such as chemical pollutant removal [29,30], chromatographic separation [31,32], and drug delivery [33]. ZIF-67 is a recently developed ZIF compound that has the formula $Co(Hmim)_2$ (mim = 2-methylimidazole) with a sodalite-related zeolite type structure [34]. Because of the low coordination of the Co cation, ZIF-67 has three times higher adsorption capacity for dye from aqueous solutions than ZIF-8 [35]. Owing to the hydrogen bonding and π–π electron donor–acceptor interactions between the adsorbent and analytes, Fe_3O_4–MWCNTs–OH@poly-ZIF67 shows good selective extraction of aromatic acids [36]. This research indicates that magnetic nanocarbon-material-based ZIFs have good application prospects for adsorbing pesticides because of their good adsorptive properties derived from the magnetic carbon nanocomposite.

Organochlorine pesticides (OCPs) are ubiquitous in the environment due to their extensive application and persistent organic pollutants characteristics, which pose great risks to human health and ecosystems [37]. On account of their bioaccumulation, degradation resistance and carcinogenesis, tetratogenesis, and mutagenesis to human, 15 OCPs, including hexachlorocyclohexane (HCHs) and dichlorodiphenyltrichloroethane (DDTs), were banned with the issue of the Stockholm Convention in 2004. However, the current study showed that the concentration level of OCPs in aqueous environment around Beijing were ranged from 9.81 to 32.1 ng L^{-1} (average 15.1 $\pm$ 7.78 ng L^{-1}) [38]. Therefore, it is necessary to develop sensitive and accuracy analytical methods for continuous monitoring trace level of the OCPs in water samples.

Inspired by the abovementioned studies, a novel magnetic Co-based ZIF composite was synthesized by organic–inorganic coordination. The obtained M-M-ZIF-67 composite possessed porous surfaces and super-paramagnetism due to the utilization of Fe_3O_4–MWCNTs as the magnetic kernel and support. In addition, due to the large surface area and excellent adsorption capacity of MWCNTs, the prepared hybrid material could be good adsorbent for MSPE of OCPs. In the end, an M-M-ZIF-67 based MSPE method was established and applied for the extraction of OCPs from agricultural irrigation water samples prior to gas chromatography–tandem triple quadrupole mass spectrometry (GC–MS/MS).

2. Materials and Methods

2.1. Reagents and Materials

The standard liquid pesticides α-HCH (CAS number: 319-84-6), β-HCH (CAS number: 319-85-7), γ-HCH (CAS number: 58-89-9), δ-HCH (CAS number: 319-86-8), p,p'-DDD (CAS number: 72-54-8), o,p-DDE (CAS number: 3424-82-6), p,p'-DDE (CAS number: 72-55-9), o,p-DDT (CAS number: 789-02-6), and p,p'-DDT (CAS number: 50-29-3) were obtained from the Agro-Environmental Protection Institute, Ministry of Agriculture (Tianjin, China) at concentrations of 1000 mg L^{-1}. A standard mixture containing 20 mg L^{-1} of each of the nine OCPs was prepared in methanol and stored at −20 °C in the dark. High-performance liquid chromatography grade acetonitrile, methanol, and n-hexane were purchased from Sigma-Aldrich (St. Louis, MO, USA). Anhydrous sodium sulfate was supplied by Agilent (CA, California, USA). The MWCNTs (8–15 nm inner diameter (id), 10–30 μm long, 95% purity), analytical grade ferric chloride hexahydrate ($FeCl_3 \bullet 6H_2O$), ferrous chloride tetrahydrate ($FeCl_2 \bullet 4H_2O$), 2-methylimidazole, cobalt nitrate hexahydrate ($Co(NO_3)_2 \bullet 6H_2O$), and ammonium hydroxide (mass fraction 28%) were provided by Aladdin Co. (Shanghai, China). Ethanol and all of the other reagents were of analytical grade and acquired from the Beijing Chemical Reagents Co. (Beijing, China).

2.2. Preparation of Fe_3O_4–MWCNTs–ZIF-67

2.2.1. Synthesis of Fe_3O_4–MWCNTs

The Fe_3O_4–MWCNTs were prepared according to the authors' previously reported method with slight modification [39]. In brief, MWCNT powder (0.2 g) was suspended in ultrapure water (240 mL) by sonication for 1 h and then transferred to a three-necked flask. After a solution of $FeCl_3 \bullet 6H_2O$ (1.8 g) and $FeCl_2 \bullet 4H_2O$ (0.8 g) dissolved in ultrapure water (25 mL) was added to the flask, the mixture was vigorously stirred with a mechanical stirrer (THZ-82A, Youlian instrument research institute, Jintan, China) under protection of N_2 at 150 rpm and 80 °C conditions for 30 min. Ammonium hydroxide (28%, 10 mL) was then added and the mixture was vigorously stirred for another 30 min. After cooling to room temperature, the sediments were collected by magnetic separation and washed three times with ethanol and ultrapure water to eliminate unreacted chemicals. The obtained Fe_3O_4–MWCNT nanoparticles were dried in a vacuum oven at 60 °C for 24 h.

2.2.2. Synthesis of ZIF-67 and M-M-ZIF-67

The ZIF-67 material was fabricated following a reported method [40]. The preparation procedure for M-M-ZIF-67 was as follows. First, the Fe_3O_4–MWCNTs were dispersed in ultrapure water (20 mL) and mixed with 3 mL of an aqueous solution of $Co(NO_3)_2 \bullet 6H_2O$ (0.45 g) with consistent stirring for 30 min under 150 rpm condition. An aqueous solution of 2-methylimidazole (20 mL, 0.45 g) was then added to the solution and the solution was stirred for 6 h. All of these synthetic processes were performed at room temperature. Finally, the M-M-ZIF-67 product was obtained by magnetic separation and washed three times with ethanol and ultrapure water. The synthesized material was dried in a vacuum oven at 60 °C for 24 h.

2.3. MSPE Procedure

The workflow of MSPE using M-M-ZIF-67 is shown in Figure 1. First, M-M-ZIF-67 (6 mg) was placed in a 10 mL centrifuge tube containing 5.0 mL of the aqueous standard solution or sample solution and shaken for 20 min for extraction. With aggregation of the adsorbent in the tube, the supernatant was discarded with the aid of an external ferrite magnet. Acetonitrile (2 mL) was then added into the tube, and ultrasonic elution of the analytes from the magnetic materials was performed for 5 min. After the M-M-ZIF-67 composite was collected, the supernatant desorption solution was transferred to another centrifuge tube. The same desorption procedures were performed one more time. Finally, the combined desorbed elution was evaporated to dryness under a gentle stream of nitrogen at 40 °C. The residue was redissolved in 0.5 mL acetone, and 1 μL of it was injected into the GC-MS/MS for analysis.

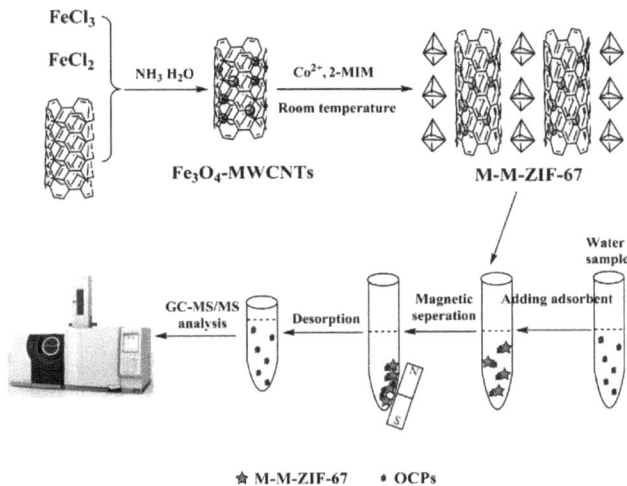

Figure 1. Schematic illustration of the synthetic route to prepare zeolitic imidazolate framework based on magnetic multiwalled carbon nanotubes and the magnetic solid-phase extraction steps for organochlorine pesticides analysis.

2.4. Sample Preparation

The river water sample was collected from the Liangshui River, Beijing, China. The tap water sample was obtained from the tap in the laboratory; and the underground water sample was obtained from Langfang City, Hebei Province, China. All of the samples were filtered through a 0.45 μm polytetrafluoroethylene membrane filter and stored at 4 °C in amber dark glass bottles.

2.5. Apparatus and Gas Chromatography–Tandem Triple Quadrupole Mass Spectrometry Conditions

The surface morphologies and particle sizes of the as-synthesized nanoparticles were observed by scanning electron microscopy (SEM, JSM-6300, JEOL, Tokyo, Japan) and transmission electron microscopy (TEM, JEM-200CX, JEOL, Tokyo, Japan). The powder X-ray diffraction (XRD) patterns of the as-synthesized nanoparticles were obtained with an X-ray powder diffractometer (D8 Advance, Bruker, Karlsruhe, Germany). The Fourier-transform infrared (FT-IR) spectra of the as-synthesized nanoparticles were obtained with an FT-IR-8400 spectrometer (Shimadzu, Kyoto, Japan). A vibrating sample magnetometer (VSM, Lake Shore 7410, Columbus, OH, USA) was used to investigate the magnetic properties of all of the synthetic materials.

Gas chromatography–tandem triple quadrupole mass spectrometry (GC–MS/MS) analysis was performed with a Shimadzu GC-2010 plus gas chromatograph coupled with an AOC-20s autosampler, a Shimadzu TQ8040 triple-quadrupole MS. The pesticides were separated on an Rtx-5MS capillary column purchased from RESTEK (Bellefonte, PA, USA) (0.25 mm (id) × 30 m, 0.25 μm film thickness). Helium gas was used as the carrier gas at a constant flow rate of 1 mL min^{-1}. The column temperature was programmed as follows: the initial temperature of 40 °C was maintained for 4 min, the temperature was increased to 125 °C at 25 °C min^{-1}, the temperature was ramped to 300 °C at 10 °C min^{-1}, and the temperature was maintained at 300 °C for 6 min. The total run time was 30.9 min. The injector temperature was 250 °C, and the injection volume was 1.0 μL in splitless mode. The specific multiple reaction monitoring (MRM) transitions for all the nine OCPs and the other parameters are given in Table 1.

Table 1. Acquisition and chromatographic parameters of the nine organochlorine pesticides.

Pesticides	Retention Time (min)	MRM1 [a] (*m/z*)	CE1 [b] (eV)	MRM2 (*m/z*)	CE2 (eV)
α-HCH [c]	15.32	218.90 > 182.90	8	218.90 > 144.90	20
β-HCH	15.87	218.90 > 182.90	8	218.90 > 144.90	20
γ-HCH	16.01	218.90 > 182.90	8	218.90 > 144.90	20
δ-HCH	16.59	218.90 > 182.90	10	218.90 >144.90	20
o,p′-DDE [d]	19.47	246.00 > 176.00	30	246.00 > 211.00	22
p,p′-DDE	20.09	246.00 > 176.00	30	246.00 > 211.00	22
p,p′-DDD [e]	20.90	235.00 > 165.00	24	235.00 > 199.00	14
o,p′-DDT [f]	20.95	235.00 > 165.00	24	235.00 > 199.00	16
p,p′-DDT	21.61	235.00 > 165.00	24	235.00 > 199.00	16

[a] MRM means multiple reaction monitoring transitions; [b] CE means collision energy; [c] HCH means hexachlorocyclohexane; [d] DDE means 1,1-dichloro-2,2-bis(*p*-chlorophenyl)ethylene; [e] DDD means [1,1-dichloro-2, 2-bis(*p*-chlorophenyl)ethylene; [f] DDT means 1,1,1-trichloro-2,2-bis(*p*-chlorophenyl)ethylene.

2.6. Quality Control and Quality Assurance

Quality control and quality assurance (QA/QC) experiments, including a blank sample test, recovery test, repeatability test, and limits of detection (LOD) experiment were performed to evaluate the feasibility of the method. In addition, 5 L ultrapure water served as the water sample for the blank test, while the water samples for the recovery test and repeatability test were made by spiking 50 μL 20 mg L^{-1} working solution into 100 mL ultrapure water. The LOD experiment was undertaken following the USA Environmental Protection Agency method.

3. Results

3.1. Characterization of M-M-ZIF-67

The micro-morphologies of the Fe$_3$O$_4$–MWCNTs and M-M-ZIF-67 were observed by SEM and TEM. As shown in Figure 2A, the Fe$_3$O$_4$ nanoparticles are attached to the MWCNT surface. The SEM image of M-M-ZIF-67 in Figure 2B shows that the composite has a rough surface, indicating that the material has good potential as an adsorbent [27].

VSM was performed to investigate the magnetic behavior of the magnetic materials, and the results are shown in Figure 3A. The magnetic hysteresis loops show that both the remanence and coercivity values of the three types of magnetic materials are zero, which indicates that they have typical supermagnetic properties and could be separated using an external magnet. The saturation magnetization values of Fe$_3$O$_4$, Fe$_3$O$_4$–MWCNTs, and M-M-ZIF-67 are 66.8, 59.6, and 53.1 emu g^{-1}, respectively. As shown in the insert of Figure 3A, well-dispersed M-M-ZIF-67 particles exist in the absence of an external magnet, and they are rapidly attracted to the walls of the vial in a short time (about 20 s) with application of a magnet. The powder XRD patterns of Fe$_3$O$_4$, Fe$_3$O$_4$–MWCNTs, and M-M-ZIF-67 are shown in Figure 3B. The diffraction patterns of M-M-ZIF-67 are in very close agreement with the materials of Fe_3O_4 and Fe_3O_4–MWCNTs. This indicates that Fe_3O_4 is well retained in the Fe_3O_4–MWCNTs and M-M-ZIF-67. The FT-IR spectra are shown in Figure 3C. For M-M-ZIF-67, the adsorption band at 577 cm^{-1} corresponds to the stretching vibrations of Fe–O, and the band at 1566 cm^{-1} corresponds to in-plane bending of CH_2 of the MWCNTs. This suggests that M-M-ZIF-67 is composed of Fe_3O_4 and MWCNTs. The bands in the region 673 to 1454 cm^{-1} are attributed to complete ring stretching or bending vibration of 2-methylimidazole, and the main adsorption bands are preserved in the spectrum of the composite, which indicates that ZIF-67 is successfully synthesized on the surface of M-M-ZIF-67.

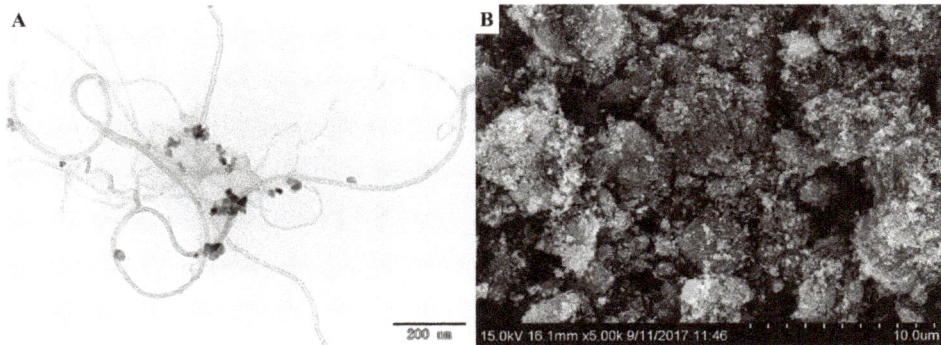

Figure 2. (**A**) transmission electron microscopy image of magnetic multiwalled carbon nanotubes and (**B**) scanning electron microscopy image of zeolitic imidazolate framework based on magnetic multiwalled carbon nanotubes.

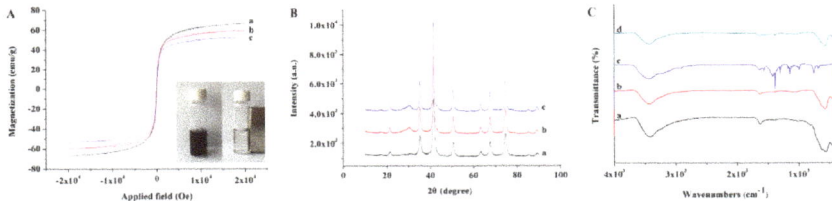

Figure 3. (**A**) magnetic hysteresis loops for (**a**) Fe_3O_4; (**b**) magnetic multiwalled carbon nanotubes; and **c**) zeolitic imidazolate framework based on magnetic multiwalled carbon nanotubes. The insert shows magnetic separation and dispersion of zeolitic imidazolate framework based on magnetic multiwalled carbon nanotubes; (**B**) X-ray diffraction patterns of (**a**) Fe_3O_4; (**b**) magnetic multiwalled carbon nanotubes; and **c**) zeolitic imidazolate framework based on magnetic multiwalled carbon nanotubes; (**C**) Fourier-transform infrared spectra of (**a**) Fe_3O_4; (**b**) magnetic multiwalled carbon nanotubes; (**c**) zeolitic imidazolate framework-67; and (**d**) zeolitic imidazolate framework based on magnetic multiwalled carbon nanotubes.

3.2. Optimization of the MSPE Parameters

The MSPE parameters affect extraction of the OCPs and the desorption performance. The amount of M-M-ZIF-67, extraction time, type of desorption solvent, desorption time, and frequency of desorption were investigated by single-factor experiments following a step by step procedure at a spiked level of 50 $\mu g\ L^{-1}$.

3.2.1. Optimization of the Extraction Process

The present study aims to achieve satisfactory extraction performance for analytes in water samples. The effect of the amount of M-M-ZIF-67 on extraction was investigated using different amounts of the sorbent ranging from 2 to 10 mg. The experimental results (Figure 4A) show that the extraction recoveries of the OCPs rapidly increase when the amount of M-M-ZIF-67 is increased from 2 to 6 mg, and the recoveries then slightly decrease when the amount of sorbent is increased from 6 to 10 mg. Based on these results, 6 mg of M-M-ZIF-67 was chosen for the following experiments.

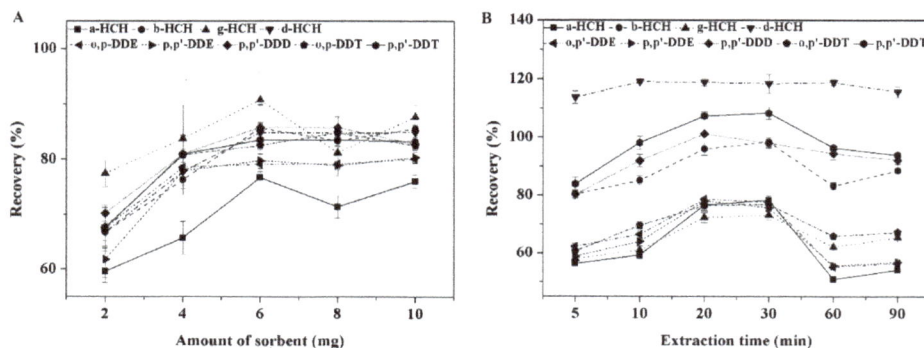

Figure 4. Optimization of the extraction process for 50 µg L^{-1} organochlorine pesticides using zeolitic imidazolate framework based on magnetic multiwalled carbon nanotubes: (**A**) amount of sorbent and (**B**) extraction time. Desorption conditions: 2 mL of acetonitrile; ultrasonication time, 5 min; and ultrasonication performed one more time. The error bars show the standard deviation of the mean (n = 3).

The extraction time is an important factor for MSPE because it influences the adsorption equilibrium of the analytes between the sample solution and adsorbent. Therefore, the extraction time was varied in the range 5–90 min to investigate its influence on the extraction efficiency. As shown in Figure 4B, the recoveries of the target OCPs increase with increasing extraction time from 5 to 20 min, and they then remain almost constant until 30 min. The extraction recoveries for most of the analytes then decrease with a further increase of the extraction time. This is probably because of desorption of the analytes from the adsorbent, especially when the adsorption equilibrium is reached. Thus, 20 min was chosen as the extraction time for the following experiments.

3.2.2. Optimization of the Desorption Process

The desorption solvent is crucial and it can significantly affect the MSPE efficiency. Three different solvents were investigated as desorption solvents to investigate their effect on the extraction efficiency: acetonitrile, acetone, and n-hexane. Figure 5A shows that the extraction recoveries of the target analytes eluted by acetonitrile are better than those with the other solvents. Therefore, acetonitrile was used as the desorption solvent for the following experiments.

The desorption time is another important factor that influences the efficiency of the MSPE process. To investigate the influence of the desorption time on the MSPE efficiency, experiments were performed with ultrasonic desorption times of 2, 5, 10, and 15 min. As shown in Figure 5B, the extraction recoveries of the analytes are satisfactory when the sample solution is ultrasonicated for 5 and 10 min. Considering the efficiency, an ultrasonic desorption time of 5 min was chosen for the subsequent experiments.

To investigate the effect of the frequency of desorption, ultrasonic desorption of the nine OCPs was performed one to four times using 2 mL of the desorption solvent and a desorption time of 5 min.

The results are shown in Figure 5C. The optimum recoveries are obtained when the analytes are eluted two times. Thus, the frequency of desorption was set to two for the following experiments.

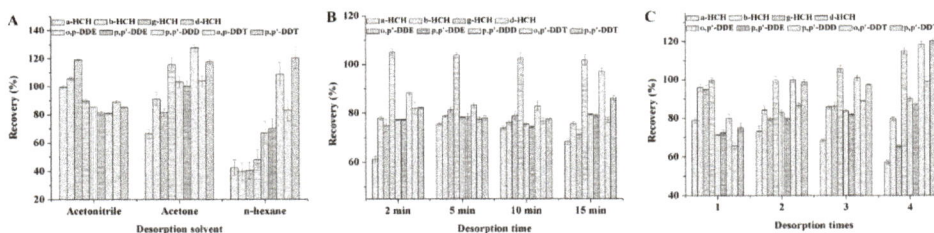

Figure 5. Optimization of the desorption process for 50 μg L^{-1} organochlorine pesticides using zeolitic imidazolate framework based on magnetic multiwalled carbon nanotubes: (**A**) desorption solvent; (**B**) desorption time; and (**C**) frequency of desorption. Extraction conditions: sample volume, 5 mL; amount of sorbent, 6 mg; and extraction time, 20 min. The error bars show the standard deviation of the mean (*n* = 3).

3.3. Method Characterization

A series of experiments was performed under the optimized conditions (Table 2) and the analytical characteristics, such as the linear ranges (LRs), LOD, and relative standard deviations (RSDs), were determined to validate the developed method. Working solutions containing the target analytes at concentrations of 1, 2, 5, 10, 20, 50, 100, and 200 μg L^{-1} were prepared to construct the working curves. Good linearities for the nine OCPs were achieved with determination coefficients (R^2) ranging from 0.9916 to 0.9989. The LOD for all of the analytes were calculated at signal/noise (S/N) ratios of 3. The LODs range from 0.07 to 1.03 μg L^{-1}. The repeatability of the method was evaluated by performing six replicate analyses of spiked samples with 10 μg L^{-1} of each of the OCPs, and the RSDs are in the range 1.0–8.5%. All of these results indicate that the proposed method has high sensitivity and good repeatability.

Table 2. Analytical parameters of zeolitic imidazolate framework based on magnetic multiwalled carbon nanotubes as an adsorbent for magnetic solid-phase extraction of nine organochlorine pesticides in an ultrapure water sample.

OCPs	Linear Range (μg L^{-1})	Determination Coefficients	LOD [a] (μg L^{-1})	RSD [b] (%) (*n* = 6)
α-HCH	2–200	0.9953	0.12	8.5
β-HCH	1–200	0.9946	0.13	1.0
γ-HCH	1–200	0.9919	0.15	6.2
δ-HCH	1–200	0.9947	0.07	4.0
o,p′-DDE	1–200	0.9951	0.17	1.5
p,p′-DDE	1–200	0.9989	0.45	1.3
p,p′-DDD	2–200	0.9934	0.41	3.1
o,p′-DDT	2–200	0.9932	0.74	3.9
p,p′-DDT	2–200	0.9916	1.03	3.1

[a] LOD means limit of determination; [b] RSDs means relative standard deviations, and were determined by performing six replicate analyses of the spiked samples with 10 μg L^{-1} of each of the organochlorine pesticides.

3.4. Comparison

To investigate the extraction performance of the prepared materials, 4 mg of Fe$_3$O$_4$–MWCNTs and 6 mg of M-M-ZIF-67, which was synthesized based on the same amount of the former material, were applied to extraction of the nine OCPs using the proposed method. The results are shown in Figure 6 based on three replicate analyses (*n* = 3). The peak areas of all of the analytes for M-M-ZIF-67 are 2.7–3.0 times higher than those for Fe_3O_4–MWCNTs, indicating that ZIF-67 plays a greater role in the extraction process.

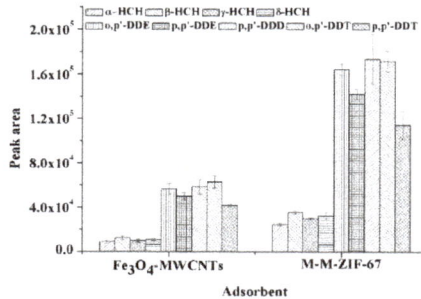

Figure 6. Effect of the sorbent on the adsorption capacities for the nine organochlorine pesticides. Extraction conditions: sample volume, 5 mL; amount of sorbent, 6 mg zeolitic imidazolate framework based on magnetic multiwalled carbon nanotubes and 4 mg magnetic multiwalled carbon nanotubes; and extraction time, 20 min. Desorption conditions: 2 mL of acetonitrile; ultrasonication time, 5 min, and ultrasonication performed one more time. The error bars show the standard deviation of the mean ($n = 3$).

The performance of the current method was compared with some recently reported methods used for analysis of OCPs in water. The comparison data (Table 3) show that the proposed method based on the M-M-ZIF-67 adsorbent exhibits good sensitivity (as the LOD), less use of sample and adsorbent. However, the Fe_3O_4–MWCNTs based adsorbent can make the phase separation process easier and faster without additional centrifugation or filtration procedures, and also can avoid the time-consuming column passing operations encountered in SPE. Therefore, the developed M-M-ZIF-67 based magnetic solid-phase extraction method showed agreement in accordance with the principles of green analytical chemistry [41].

Table 3. Comparison of different methods for analysis of organochlorine pesticides.

Method	Sorbent	Sample Amount (mL)	Number of OCPs	Volume of Elution Solvent	Sorbent Amount (mg)	Extraction Time (min)	LOD	RSD (%)	Spiked Level	Ref.
MSPE-GC-MS/MS [a]	BMZIF-derived carbon [b]	10	8	Dichloromethane, 2.00 mL	6	10	0.39–0.70 ng L^{-1}	5.5–9.1	5–500 ng L^{-1}	[42]
μ-SPE-GC-MS [c]	ZnO-CF [d]	10	15	Toluene, 0.30 mL	15	30	0.19–1.64 μg L^{-1}	2.3–10.2	1–50 μg L^{-1}	[43]
PT-SPE-GC-ECD [e]	GUF-MIR [f]	1	3	Cyclohexane-ethyl acetate (9:1, *v/v*), 0.60 mL	5	-	0.24–0.66 ng g^{-1}	3.5–6.7	2.2–220 ng g^{-1}	[44]
MSPE-GC-ECD	RGO/Fe$_3$O$_4$@Au [g]	10	6	Acetonitrile, 0.25 mL	20	10	0.4–4.1 μg L^{-1}	1.7–7.3	100 μg L^{-1}	[45]
MSPE-GC-μECD	Fe$_3$O$_4$@MAA@IBL [h]	20	5	Acetonitrile, 0.20 mL	20	10	1.0–1.9 ng L^{-1}	6.2–8.3	100 μg L^{-1}	[46]
d-μSPE [i]-GC-MS	rGO-amino-HNT@PT [j]	10	6	Acetonitrile, 0.50 mL	5	5	2–13 ng L^{-1}	6.1–9.7	5–70 μg L^{-1}	[47]
μ-SPE-GC-MS	MIL-101 [k]	10	5	Ethyl acetate, 0.20 mL	4	40	2.5–16 ng L^{-1}	4.2–11.0	10 μg L^{-1}	[48]
MSPE-GC/ECD	β-CD/MRGO [l]	50	16	acetonitrile-dichloromethane (4:1, *v/v*), 1.00 mL	15	3	0.5–3.2 ng Kg^{-1}	3.3–7.8	50 ng kg^{-1}	[49]
MSPE-GC-MS/MS	M-M-ZIF-67 [m]	5	9	Acetonitrile, 4.00 mL	6	20	0.07–1.03 μg L^{-1}	1.0–8.5	10 μg L^{-1}	This work

[a] magnetic solid-phase extraction followed by gas chromatography–tandem triple quadrupole mass spectrometry determination; [b] micro solid-phase extraction followed by gas chromatography–mass spectrometry determination; [c] Bimetallic MOF; [d] Zinc oxide nanoparticles incorporated in carbon foam; [e] Miniaturized pipette tip solid-phase extraction followed by gas chromatography combination with electronic capture detector determination; [f] Clyoxal–urea–formaldehyde molecularly imprinted resin; [g] Reduced graphene oxide/Fe$_3$O$_4$@gold nanocomposite; [h] Fe$_3$O$_4$@mercaptoacetic acid@imine-based ligand; [i] dispersive micro solid-phase extraction; [j] Reduced graphene oxide-amino-halloysite nanotubes@polythiophene; [k] metal-organic framework; [l] β-Cyclodextrin/iron oxide-reduced graphene oxide hybrid nanostructure; [m] zeolitic imidazolate framework based on magnetic multiwalled carbon nanotubes.

3.5. Real Sample Analysis

The proposed MSPE method was applied to determination of OCPs in real agricultural irrigation water samples, including tap water, river water, and underground water. The results are summarized in Table 4 and the total ion chromatograms (TICs) of the analytes acquired from the tap water samples are shown in Figure 7. There are no OCPs in the selected water samples and the recovery results indicate that the developed method has good utility for analysis of OCPs in real water samples.

Table 4. Analytical results for determination of organochlorine pesticides in real water samples.

Matrix	Analyte	Spiked Concentration ($\mu g\ L^{-1}$, $n = 3$)				
		0	10		100	
		Found	Recovery (%)	RSD (%)	Recovery (%)	RSD (%)
	α-HCH	<LOD	83.4	7.4	84.7	0.7
	β-HCH	<LOD	92.5	9.6	103.7	0.6
	γ-HCH	<LOD	93.3	5.5	91.9	0.4
	δ-HCH	<LOD	94.6	6.6	111.1	1.0
Tap water	o,p'-DDE	<LOD	85.4	3.4	99.6	5.3
	p,p'-DDE	<LOD	78.1	1.5	107.5	3.1
	p,p'-DDD	<LOD	76.0	2.5	97.0	2.5
	o,p'-DDT	<LOD	83.1	5.4	89.2	2.1
	p,p'-DDT	<LOD	93.7	2.1	105.7	1.6
	α-HCH	<LOD	76.8	5.7	80.5	2.4
	β-HCH	<LOD	91.3	3.1	102.5	0.8
	γ-HCH	<LOD	84.1	8.5	91.2	3.5
	δ-HCH	<LOD	74.9	5.9	112.7	2.6
River water	o,p'-DDE	<LOD	110.8	4.7	94.3	0.1
	p,p'-DDE	<LOD	102.5	4.8	100.9	1.2
	p,p'-DDD	<LOD	101.9	3.4	92.2	0.7
	o,p'-DDT	<LOD	108.6	4.3	86.9	1.2
	p,p'-DDT	<LOD	110.8	3.5	100.9	0.8
	α-HCH	<LOD	81.0	8.7	80.3	3.3
	β-HCH	<LOD	83.9	7.8	97.2	2.1
	γ-HCH	<LOD	90.1	7.5	80.4	5.2
	δ-HCH	<LOD	79.5	15.3	112.7	2.0
Underground water	o,p'-DDE	<LOD	101.5	8.4	82.9	1.9
	p,p'-DDE	<LOD	95.3	7.6	81.0	1.4
	p,p'-DDD	<LOD	92.8	8.9	75.1	2.2
	o,p'-DDT	<LOD	111.1	6.4	110.3	3.2
	p,p'-DDT	<LOD	116.3	3.7	90.5	2.3

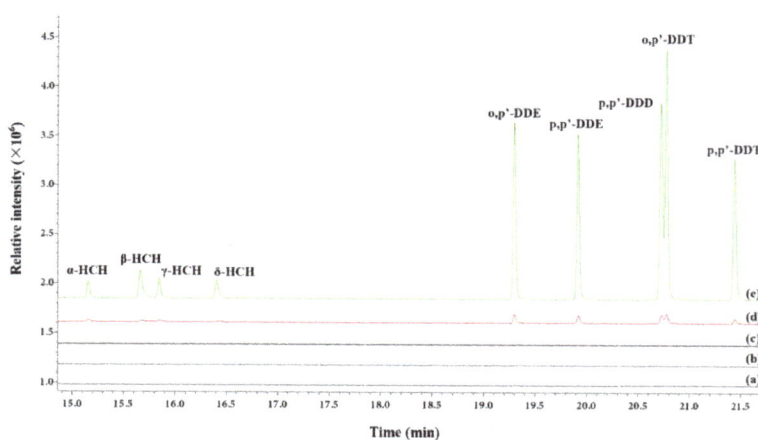

Figure 7. Total ion chromatograms of the nine organochlorine pesticides in water samples obtained by gas chromatography–tandem triple quadrupole mass spectrometry: (**a**) tap water; (**b**) river water; (**c**) underground water; (**d**) tap water spiked at 10 $\mu g\ L^{-1}$; and (**e**) tap water spiked at 100 $\mu g\ L^{-1}$.

Appl. Sci. **2018**, *8*, 959

4. Conclusions

Magnetic composites containing Fe_3O_4, MWCNTs, and ZIF-67 have been synthesized. The synthesized materials have porous surfaces and exhibit super-paramagnetism. They were used as sorbents for MSPE to extract nine OCPs from agricultural irrigation water samples. For M-M-ZIF-67, the OCP adsorption capacities are nearly three times higher than those for Fe_3O_4–MWCNTs. The developed method achieves high extraction efficiencies, good linearities, low detection limits, and good accuracies and precision. The results suggest that M-M-ZIF-67 is a simple and effective potential adsorbent for removal of OCPs from environmental water samples.

Author Contributions: X.H. and G.L. conceived and designed the experiments; X.X. and L.L. performed the experiments; X.H. and S.Z. analyzed the data; D.X. and H.L. contributed reagents and materials; X.H. and H.G. wrote the paper. Authorship must be limited to those who have contributed substantially to the work reported.

Acknowledgments: This work was supported by the National Key Research Development Program of China (Grant No. 2016YFD0200200), the National Natural Science Foundation of China (Grant No. 31701695) and the project of risk assessment on vegetable products (Grant No. GJFP2018002).

Conflicts of Interest: The authors declare no conflict of interest.

References

1. Płotka-Wasylka, J.; Szczepańska, N.; de la Guardia, M.; Namieśnik, J. Modern trends in solid phase extraction: New sorbent media. *Trends Anal. Chem.* **2016**, *77*, 23–43. [CrossRef]
2. Arthur, C.L.; Pawliszyn, J. Solid phase microextraction with thermal desorption using fused silica optical fibers. *Anal. Chem.* **1990**, *62*, 2145–2148. [CrossRef]
3. Basheer, C.; Alnedhary, A.A.; Rao, B.S.M.; Valliyaveettil, S.; Lee, H.K. Development and application of porous membrane-protected carbon nanotube micro-solid-phase extraction combined with gas chromatography/mass spectrometry. *Anal. Chem.* **2006**, *78*, 2853–2858. [CrossRef] [PubMed]
4. Sandra, P.; Tienpont, B.; David, F. Multi-residue screening of pesticides in vegetables, fruits and baby food by stir bar sorptive extraction-thermal desorption-capillary gas chromatography-mass spectrometry. *J. Chromatogr. A* **2003**, *1000*, 299–309. [CrossRef]
5. Maya, F.; Palomino Cabello, C.; Frizzarin, R.M.; Estela, J.M.; Turnes Palomino, G.; Cerdà, V. Magnetic solid-phase extraction using metal-organic frameworks (MOFs) and their derived carbons. *Trend. Anal. Chem.* **2017**, *90*, 142–152. [CrossRef]
6. Li, N.; Jiang, H.-L.; Wang, X.; Wang, X.; Xu, G.; Zhang, B.; Wang, L.; Zhao, R.-S.; Lin, J.-M. Recent advances in graphene-based magnetic composites for magnetic solid-phase extraction. *Trend. Anal. Chem.* **2018**, *102*, 60–74. [CrossRef]
7. Gałuszka, A.; Migaszewski, Z.M.; Konieczka, P.; Namieśnik, J. Analytical eco-scale for assessing the greenness of analytical procedures. *Trend. Anal. Chem.* **2012**, *37*, 61–72. [CrossRef]
8. Herrero-Latorre, C.; Barciela-García, J.; García-Martín, S.; Peña-Crecente, R.M.; Otárola-Jiménez, J. Magnetic solid-phase extraction using carbon nanotubes as sorbents: A review. *Anal. Chim. Acta* **2015**, *892*, 10–26. [CrossRef] [PubMed]
9. Guo, H.; Xue, L.; Yao, S.; Cai, X.; Qian, J. Rhein functionalized magnetic chitosan as a selective solid phase extraction for determination isoflavones in soymilk. *Carbohydr. Polym.* **2017**, *165*, 96–102. [CrossRef] [PubMed]
10. Shi, Y.; Wu, H.; Wang, C.; Guo, X.; Du, J.; Du, L. Determination of polycyclic aromatic hydrocarbons in coffee and tea samples by magnetic solid-phase extraction coupled with HPLC-FLD. *Food Chem.* **2016**, *199*, 75–80. [CrossRef] [PubMed]
11. Yilmaz, E.; Alosmanov, R.M.; Soylak, M. Magnetic solid phase extraction of lead(ii) and cadmium(ii) on a magnetic phosphorus-containing polymer (M-PhCP) for their microsampling flame atomic absorption spectrometric determinations. *RSC Adv.* **2015**, *5*, 33801–33808. [CrossRef]
12. Ma, S.; He, M.; Chen, B.; Deng, W.; Zheng, Q.; Hu, B. Magnetic solid phase extraction coupled with inductively coupled plasma mass spectrometry for the speciation of mercury in environmental water and human hair samples. *Talanta* **2016**, *146*, 93–99. [CrossRef] [PubMed]

13. Habila, M.A.; Alothman, Z.A.; El-Toni, A.M.; Labis, J.P.; Soylak, M. Synthesis and application of $Fe_3O_4@SiO_2@TiO_2$ for photocatalytic decomposition of organic matrix simultaneously with magnetic solid phase extraction of heavy metals prior to ICP-MS analysis. *Talanta* **2016**, *154*, 539–547. [CrossRef] [PubMed]

14. Su, X.; Li, X.; Li, J.; Liu, M.; Lei, F.; Tan, X.; Li, P.; Luo, W. Synthesis and characterization of core-shell magnetic molecularly imprinted polymers for solid-phase extraction and determination of Rhodamine B in food. *Food Chem.* **2015**, *171*, 292–297. [CrossRef] [PubMed]

15. Reis, L.C.; Vidal, L.; Canals, A. Graphene oxide/Fe_3O_4 as sorbent for magnetic solid-phase extraction coupled with liquid chromatography to determine 2, 4, 6-trinitrotoluene in water samples. *Anal. Bioanal. Chem.* **2017**, *409*, 2665–2674. [CrossRef] [PubMed]

16. Liu, X.; Xie, S.; Ni, T.; Chen, D.; Wang, X.; Pan, Y.; Wang, Y.; Huang, L.; Cheng, G.; Qu, W.; et al. Magnetic solid-phase extraction based on carbon nanotubes for the determination of polyether antibiotic and s-triazine drug residues in animal food with LC-MS/MS. *J. Sep. Sci.* **2017**, *40*, 2416–2430. [CrossRef] [PubMed]

17. Zhang, B.-T.; Zheng, X.; Li, H.-F.; Lin, J.-M. Application of carbon-based nanomaterials in sample preparation: A review. *Anal. Chim. Acta* **2013**, *784*, 1–17. [CrossRef] [PubMed]

18. Xu, Y.; Ding, J.; Chen, H.; Zhao, Q.; Hou, J.; Yan, J.; Wang, H.; Ding, L.; Ren, N. Fast determination of sulfonamides from egg samples using magnetic multiwalled carbon nanotubes as adsorbents followed by liquid chromatography-tandem mass spectrometry. *Food Chem.* **2013**, *140*, 83–90. [CrossRef] [PubMed]

19. Ding, J.; Gao, Q.; Li, X.-S.; Huang, W.; Shi, Z.-G.; Feng, Y.-Q. Magnetic solid-phase extraction based on magnetic carbon nanotube for the determination of estrogens in milk. *J. Sep. Sci.* **2011**, *34*, 2498–2504. [CrossRef] [PubMed]

20. Moreno, V.; Zougagh, M.; Ríos, Á. Hybrid nanoparticles based on magnetic multiwalled carbon nanotube-nano$C_{18}SiO_2$ composites for solid phase extraction of mycotoxins prior to their determination by LC-MS. *Microchim. Acta* **2016**, *183*, 871–880. [CrossRef]

21. Khan, M.; Yilmaz, E.; Soylak, M. Vortex assisted magnetic solid phase extraction of lead (ii) and cobalt (ii) on silica coated magnetic multiwalled carbon nanotubes impregnated with 1-(2-pyridylazo)-2-naphthol. *J. Mol. Liq.* **2016**, *224*, 639–647. [CrossRef]

22. Zhao, Q.; Wei, F.; Luo, Y.-B.; Ding, J.; Xiao, N.; Feng, Y.-Q. Rapid magnetic solid-phase extraction based on magnetic multiwalled carbon nanotubes for the determination of polycyclic aromatic hydrocarbons in edible oils. *J. Agric. Food Chem.* **2011**, *59*, 12794–12800. [CrossRef] [PubMed]

23. Deng, X.; Guo, Q.; Chen, X.; Xue, T.; Wang, H.; Yao, P. Rapid and effective sample clean-up based on magnetic multiwalled carbon nanotubes for the determination of pesticide residues in tea by gas chromatography–mass spectrometry. *Food Chem.* **2014**, *145*, 853–858. [CrossRef] [PubMed]

24. Ma, J.; Jiang, L.; Wu, G.; Xia, Y.; Lu, W.; Li, J.; Chen, L. Determination of six sulfonylurea herbicides in environmental water samples by magnetic solid-phase extraction using multi-walled carbon nanotubes as adsorbents coupled with high-performance liquid chromatography. *J. Chromatogr. A* **2016**, *1466*, 12–20. [CrossRef] [PubMed]

25. Rocío-Bautista, P.; González-Hernández, P.; Pino, V.; Pasán, J.; Afonso, A.M. Metal-organic frameworks as novel sorbents in dispersive-based microextraction approaches. *Trend. Anal. Chem.* **2017**, *90*, 114–134. [CrossRef]

26. Rocío-Bautista, P.; Pacheco-Fernández, I.; Pasán, J.; Pino, V. Are metal-organic frameworks able to provide a new generation of solid-phase microextraction coatings?—A review. *Anal. Chim. Acta* **2016**, *939*, 26–41. [CrossRef] [PubMed]

27. Hashemi, B.; Zohrabi, P.; Raza, N.; Kim, K.-H. Metal-organic frameworks as advanced sorbents for the extraction and determination of pollutants from environmental, biological, and food media. *Trend. Anal. Chem.* **2017**, *97*, 65–82. [CrossRef]

28. Zhao, M.; Xie, Y.; Chen, H.; Deng, C. Efficient extraction of low-abundance peptides from digested proteins and simultaneous exclusion of large-sized proteins with novel hydrophilic magnetic zeolitic imidazolate frameworks. *Talanta* **2017**, *167*, 392–397. [CrossRef] [PubMed]

29. Jiao, C.; Li, M.; Ma, R.; Wang, C.; Wu, Q.; Wang, Z. Preparation of a Co-doped hierarchically porous carbon from Co/Zn-ZIF: An efficient adsorbent for the extraction of trizine herbicides from environment water and white gourd samples. *Talanta* **2016**, *152*, 321–328. [CrossRef] [PubMed]

30. Zhang, S.; Yao, W.; Ying, J.; Zhao, H. Polydopamine-reinforced magnetization of zeolitic imidazolate framework ZIF-7 for magnetic solid-phase extraction of polycyclic aromatic hydrocarbons from the air-water environment. *J. Chromatogr. A* **2016**, *1452*, 18–26. [CrossRef] [PubMed]

31. Fu, Y.Y.; Yang, C.X.; Yan, X.P. Fabrication of ZIF-8@SiO$_2$ core-shell microspheres as the stationary phase for high-performance liquid chromatography. *Chem-Eur. J.* **2013**, *19*, 13484–13491. [CrossRef] [PubMed]

32. Srivastava, M.; Roy, P.K.; Ramanan, A. Hydrolytically stable ZIF-8@PDMS core–shell microspheres for gas-solid chromatographic separation. *RSC Adv.* **2016**, *6*, 13426–13432. [CrossRef]

33. Fang, J.; Yang, Y.; Xiao, W.; Zheng, B.; Lv, Y.-B.; Liu, X.-L.; Ding, J. Extremely low frequency alternating magnetic field–triggered and MRI–traced drug delivery by optimized magnetic zeolitic imidazolate framework-90 nanoparticles. *Nanoscale* **2016**, *8*, 3259–3263. [CrossRef] [PubMed]

34. Zhou, Q.; Zhu, L.; Xia, X.; Tang, H. The water-resistant zeolite imidazolate framework 67 is a viable solid phase sorbent for fluoroquinolones while efficiently excluding macromolecules. *Microchim. Acta* **2016**, *183*, 1839–1846. [CrossRef]

35. Kang, X.-Z.; Song, Z.-W.; Shi, Q.; Dong, J.-X. Utilization of zeolite imidazolate framework as an adsorbent for the removal of dye from aqueous solution. *Asian J. Chem.* **2013**, *25*, 8324–8328.

36. Li, W.-K.; Chen, J.; Zhang, H.-X.; Shi, Y.-P. Selective determination of aromatic acids by new magnetic hydroxylated MWCNTs and MOFs based composite. *Talanta* **2017**, *168*, 136–145. [CrossRef] [PubMed]

37. Rani, M.; Shanker, U.; Jassal, V. Recent strategies for removal and degradation of persistent & toxic organochlorine pesticides using nanoparticles: A review. *J. Environ. Manag.* **2017**, *190*, 208–222.

38. Shao, Y.; Han, S.; Ouyang, J.; Yang, G.; Liu, W.; Ma, L.; Luo, M.; Xu, D. Organochlorine pesticides and polychlorinated biphenyls in surface water around Beijing. *Environ. Sci. Pollut. Res.* **2016**, *23*, 24824–24833. [CrossRef] [PubMed]

39. Liu, G.; Li, L.; Xu, D.; Huang, X.; Xu, X.; Zheng, S.; Zhang, Y.; Lin, H. Metal-organic framework preparation using magnetic graphene oxide-β-cyclodextrin for neonicotinoid pesticide adsorption and removal. *Carbohydr. Polym.* **2017**, *175*, 584–591. [CrossRef] [PubMed]

40. Liu, X.; Wang, C.; Wu, Q.; Wang, Z. Metal-organic framework-templated synthesis of magnetic nanoporous carbon as an efficient absorbent for enrichment of phenylurea herbicides. *Anal. Chim. Acta* **2015**, *870*, 67–74. [CrossRef] [PubMed]

41. Płotka-Wasylka, J. A new tool for the evaluation of the analytical procedure: Green analytical procedure index. *Talanta* **2018**, *181*, 204–209. [CrossRef] [PubMed]

42. Liu, Y.; Gao, Z.; Wu, R.; Wang, Z.; Chen, X.; Chan, T.W.D. Magnetic porous carbon derived from a bimetallic metal-organic framework for magnetic solid-phase extraction of organochlorine pesticides from drinking and environmental water samples. *J. Chromatogr. A* **2017**, *1479*, 55–61. [CrossRef] [PubMed]

43. Sajid, M.; Basheer, C.; Mansha, M. Membrane protected micro-solid-phase extraction of organochlorine pesticides in milk samples using zinc oxide incorporated carbon foam as sorbent. *J. Chromatogr. A* **2016**, *1475*, 110–115. [CrossRef] [PubMed]

44. Yang, C.; Lv, T.; Yan, H.; Wu, G.; Li, H. Glyoxal-urea-formaldehyde molecularly imprinted resin as pipette tip solid-phase extraction adsorbent for selective screening of organochlorine pesticides in spinach. *J. Agric. Food Chem.* **2015**, *63*, 9650–9656. [CrossRef] [PubMed]

45. Mehdinia, A.; Rouhani, S.; Mozaffari, S. Microwave-assisted synthesis of reduced graphene oxide decorated with magnetite and gold nanoparticles, and its application to solid-phase extraction of organochlorine pesticides. *Microchim. Acta* **2016**, *183*, 1177–1185. [CrossRef]

46. Mehdinia, A.; Einollahi, S.; Jabbari, A. Magnetite nanoparticles surface-modified with a zinc (ii)-carboxylate schiff base ligand as a sorbent for solid-phase extraction of organochlorine pesticides from seawater. *Microchim. Acta* **2016**, *183*, 2615–2622. [CrossRef]

47. Darvishnejad, M.; Ebrahimzadeh, H. Halloysite nanotubes functionalized with a nanocomposite prepared from reduced graphene oxide and polythiophene as a viable sorbent for the preconcentration of six organochlorine pesticides prior to their quantitation by GC/MS. *Microchim. Acta* **2017**, *184*, 3603–3612. [CrossRef]

48. Huang, Z.; Lee, H.K. Micro-solid-phase extraction of organochlorine pesticides using porous metal-organic framework mil-101 as sorbent. *J. Chromatogr. A* **2015**, *1401*, 9–16. [CrossRef] [PubMed]

49. Mahpishanian, S.; Sereshti, H. One-step green synthesis of β-cyclodextrin/iron oxide-reduced graphene oxide nanocomposite with high supramolecular recognition capability: Application for vortex-assisted magnetic solid phase extraction of organochlorine pesticides residue from honey samples. *J. Chromatogr. A* **2017**, *1485*, 32–43. [PubMed]

applied
sciences

MDPI

Article

Adsorption of Gold(I) and Gold(III) Using Multiwalled Carbon Nanotubes

Francisco Jose Alguacil

Centro Nacional de Investigaciones Metalurgicas (CSIC), Avda. Gregorio del Amo 8, 28040 Madrid, Spain; fjalgua@cenim.csic.es

Received: 25 September 2018; Accepted: 10 October 2018; Published: 16 November 2018

Featured Application: Recovery of gold from secondary wastes.

Abstract: Carbon nanotubes are materials that have been investigated for diverse applications including the adsorption of metals. However, scarce literature has described their behavior in the case of the adsorption of precious metals. Thus, this work reports the efficient adsorption of gold from cyanide or chloride media on multiwalled carbon nanotubes (MWCNTs). In a cyanide medium, gold was adsorbed from alkaline pH values decreasing the adsorption as the pH values were increased to more acidic values. In a chloride medium, the MWCNTs were able to load the precious metal and an increased HCl concentration (0.1–10 M), in the aqueous solution, had no effect on the gold uptake onto the nanotubes. From both aqueous media, the metal adsorption was well represented by the pseudo-second order kinetic model. In the cyanide medium, the film-diffusion controlled process best fitted the rate law governing the adsorption of gold onto the nanotubes, whereas in the chloride medium, the adsorption of the metal onto the nanotubes is best represented, both at 20 °C and 60 °C, by the particle-diffusion controlled process. With respect to the elution step, in cyanide medium gold loaded onto the nanotubes can be eluted with acidic thiourea solutions, whereas in the chloride medium, and due to that the adsorption process involved the precipitation of zero valent gold onto the multiwalled carbon nanotubes, the elution has been considered as a leaching step with aqua regia. From the eluates, dissolved gold can be conveniently precipitated as zero valent gold nanoparticles.

Keywords: multiwalled carbon nanotubes; gold(I); gold(III); adsorption; elution; gold nanoparticles

1. Introduction

Nearly 25 years ago, Iijima began considering carbon nanotubes [1]. After this time, scientists began finding applications (including but not excluding: medicine, drug delivery, solar cells, electronics) for carbon nanotubes technologies in different configurations (single and multi-walled carbon nanotubes, functionalized carbon nanotubes) and for different purposes due to their properties including mechanical strength, aspect ratio and electrical and thermal conductivity. As recent literature has indicated, these nanomaterials are unconditionally considered as suitable adsorbents in the treatment of waters and in the recovery and separation of metal species from aqueous solutions [2–12].

Referring to gold adsorption, little information is available using this nanotechnology, despite the fact of the importance and price (reaching USD1244 per ounce barrier as at 16th July 2018) of this metal and its presence in aqueous solutions coming from the processing of ores (alkaline cyanide medium) or jewelry scraps and printed circuit boards (PCBs) (acidic chloride medium). The reference [13] mentioned in the manuscript title about "prerequisites of carbon nanotubes to adsorb gold(III)", though then in the text and conclusions not special mention to such requisites were given, the second reference on the topic is about the adsorption of Cr(VI) and Au(III) onto oxidized multiwalled carbon nanotubes [14].

Appl. Sci. **2018**, *8*, 2264; doi:10.3390/app8112264   113   www.mdpi.com/journal/applsci

Due to this lack of recent information, in the present investigation, the adsorption of gold from these two aqueous media (alkaline cyanide or acidic chloride) by multi-walled carbon nanotubes was considered. The aim was to optimize various operational parameters and, thus, obtain efficient carbon nanotubes processing.

2. Experimental

2.1. Reagents and Solutions

The multi-walled carbon nanotubes were obtained from Fluka and their main characteristics are summarized in Table 1. Further characterization of the carbon nanotubes, including zero potential and Micro-Raman spectra are described elsewhere [15]. Other chemicals used were AR (analytical reagent) grade, except the complex $KCu(CN)_4$, which was prepared according to the literature [16]. Gold cyanide solutions were prepared by dissolving $KAu(CN)_2$ in distilled water. The different concentrations used in the experiments were prepared from a stock solution of 0.1 g/L gold(I). The pH of the solution was adjusted by the addition of hydrochloric acid or sodium hydroxide solutions. During the experiments, the pH was continuously controlled using a 605 pH-meter (2000, Crison, Madrid, Spain). In acidic chloride media, the gold(III) solutions were prepared in a similar manner using $HAuCl_4$.

Table 1. Characteristics of the multi-walled carbon nanotubes.

Type	Multi-Walled
melting range	3652–3697 °C
density	2.1 g mL^{-1}
appearance	dust
purity	$\geq$98% carbon basis
dimensions	10 ± 1 nm external diameter
	4.5 ± 0.5 nm internal diameter
	3–6 μm (length)
maximum adsorption	$1295 \text{ cm}^3 \text{ g}^{-1}$
BET	$263 \text{ m}^2 \text{ g}^{-1}$

2.2. Gold Adsorption Measurements

The adsorption experiments were carried out using a glass reactor with mechanical (27 mm diameter four blades impeller) stirring. The desired gold solution (100 mL) was put into the reactor and to this the corresponding amount of the multi-walled carbon nanotubes was added. The elution experiments were carried out in a similar device, putting together the gold-loaded carbon nanotubes and the elution phase.

Gold or metal concentrations in the aqueous solution were analyzed by atomic absorption spectrometry, using a Perkin Elmer 1100B spectrophotometer (wavelength 242.8 nm, 1993, Perkin Elmer, Oxford, UK). The percentage of metal adsorption was calculated by the difference between the initial metal concentration and the corresponding one encountered in the solution at elapsed times.

SEM and Energy Dispersive X-ray Spectrodcopy (EDS) analyses were carried out in a Hitachi S-4800 equipment (2009, Hitachi, Osaka, Japan).

3. Results and Discussion

3.1. Carbon Nanotubes Characterization

As it is mentioned in Section 2.1, the characterization of the carbonaceous material used in the present work was described elsewhere [15], though for the benefit of the readers of the present work some data were included here. This material presented an isoelectric point of 3.6, whereas Raman data are summarized in Table 2.

Table 2. Raman data of the multiwalled carbon nanotubes.

	D	G	D′	2D	G+D
Raman shift	1339	1573	1607	2675	2913
Area signal	102.4	76.2	6.8	90.1	106.8

The I_D:I_G value obtained for these multiwalled carbon nanotubes is 1.34, indicating the presence of little defects in their structure.

3.2. Gold Adsorption from Cyanide Media

The literature remarks that the gold(I) cyanide complex is adsorbed predominantly onto activated carbon as an ion pair [17], thus similarly, at neutral-alkaline pH values, the adsorption of gold(I) cyanide by the carbon nanotubes may be broadly represented by the general equilibrium:

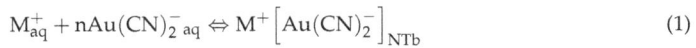

$$M^+_{aq} + nAu(CN)^-_{2\ aq} \Leftrightarrow M^+ \left[Au(CN)^-_2 \right]_{NTb} \tag{1}$$

which is, as is demonstrated below, pH-dependent; aq and NTb denote species in the aqueous and carbon phases, respectively, and M being the accompanying metal.

3.2.1. Influence of Stirring Speed of the Aqueous Phase

In all of the adsorption experiments, a stirring speed of 600 min^{-1} was used for the aqueous phases. Previous experiments, carried out with an aqueous phase containing 0.01 g L^{-1} gold(I) at pH 8 ± 0.1, and a carbon nanotube dosage of 1 g L^{-1}, showed that the maximum percentage of gold(I) removal becomes virtually independent (40 % after 2 h) of the stirring speed in the 500–1000 min^{-1} range, which also indicates that a minimum value of the thickness of the aqueous phase boundary layer is reached. Thus, a stirring speed of 600 min^{-1} was used in subsequent experiments.

3.2.2. Effect of Aqueous pH Value

Figure 1 shows the results obtained for the adsorption of Au(CN)$_2$$^-$ at various aqueous pH values. The carbon nanotubes dosage used in the experiments was of 2 g L^{-1}. The aqueous phase contained 0.01 g L^{-1} of gold(I) at various pH values. The results shown in Figure 1 demonstrate that the variation of the aqueous pH value increases the removal of gold when these pH values are shifted towards more acidic values. It should be noted here that the gold(I)-cyanide complex is stable even at acidic pH values and only decomposes at these low pH values after a long time (several hours) to yield a yellow AuCN solid. The experimental values obtained at pH 13 were used to estimate the rate law in the adsorption of gold(I) by these multiwalled carbon nanotubes, experimental data best fitted to the film-diffusion controlled mechanism (r^2: 0.9542):

$$\ln(1 - F) = -kt \tag{2}$$

where F is the factorial approach to the equilibrium, defined as:

$$F = \frac{[Au]_{nt,t}}{[Au]_{nt,e}} \tag{3}$$

where $[Au]_{nt,t}$ and $[Au]_{nt,e}$ are the gold concentration in the nanotubes at an elapsed time t and at equilibrium, respectively, and k is the rate constant (0.59 min^{-1}).

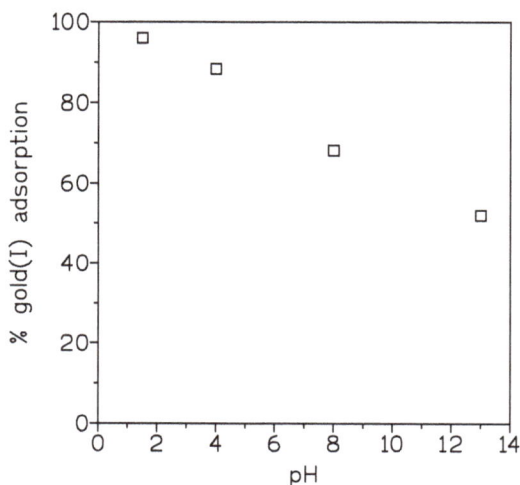

Figure 1. Effect of the pH variation on gold(I) adsorption. Time: 2 h.

3.2.3. Effect of Temperature

Aqueous solutions containing 0.01 g L^{-1} gold(I) at pH 13 ± 0.1 were used to investigate the effect of this variable on gold(I) adsorption. The adsorbent dosage was 2 g L^{-1} and the temperature was varied from 20 to 60 °C. From the results presented in Table 3, it can be shown that the increase in the temperature leads to a decrease in metal adsorption. From the D (distribution coefficient) value, defined as:

$$D = \frac{[Au]_{nt,e}}{[Au]_{s,e}} \tag{4}$$

and plotting log D versus 1/T, the slope indicated that the adsorption reaction has an exothermic character ($\Delta H° = -13$ kJ mol^{-1}) and the intercept that $\Delta S°$ is -50 J mol^{-1} K^{-1}. In the above reaction, $[Au]_{nt,e}$ and $[Au]_{s,e}$ are the gold equilibrium concentrations in the nanotubes and in the solution, respectively. From the values of the enthalpy and the entropy values, the value of $\Delta G°$ was calculated as -28 kJ mol^{-1}. The negative energy value indicated a spontaneous adsorption, whereas the negative entropy change is an indication of a decreased randomness at the solid/liquid interface during the adsorption process.

Table 3. Effect of temperature on gold(I) adsorption.

Temperature, °C	% Gold Adsorption
20	52.5
40	44.0
60	36.5

Time: 1 h.

3.2.4. Effect of the Aqueous Phase/Adsorbent Relationship on Gold Removal

In Table 4, the gold percentage of adsorption, at different aqueous phase/adsorbent relationships, is given in order to study the effect of this variation on gold(I) adsorption. Experiments were carried out at a constant pH value of 13 ± 0.1. Aqueous phases contained 0.01 g L^{-1} gold(I). The results obtained reveal an important change in the metal adsorption at lower aqueous phase/adsorbent relationships, since this adsorption is in the 90% range when the lowest relationships are used against recoveries of about 30%, obtained when the above relationship is 4000. In this same Table, the metal uptake is also given, showing the increase of this value as the ratio liquid/solid is increased.

Table 4. Influence of the liquid/solid relationship on gold(I) adsorption.

L/S Relationship	% Gold Adsorption [a]	Gold Uptake, mg g^{-1} [a]
125	93.0	1.2
250	90.0	2.3
500	57.0	2.9
1000	50.0	5.0
2000	40.5	8.2
4000	31.5	12.8

[a] After 2 h.

The values in Table 4 were used to estimate the loading isotherm, the fit of the experimental data responded best to the Langmuir isotherm but with a discrete r^2 value of 0.8338.

3.2.5. Separation of Gold (I) from Metal-Cyanide Complexes

In order to investigate the effect of other metal-cyanide complexes accompanying $Au(CN)_2^-$, a study was carried out about their interference with the overall transport of gold(I). The metal-cyanide complexes studied, $Cu(CN)_4^{3-}$, $Fe(CN)_6^{3-}$ and $Fe(CN)_6^{4-}$, were investigated in the form of binary mixtures with $Au(CN)_2-$ on a 1:1 molar basis. Thus, the feed phase contained 5.1×10^{-5} M of each metal in aqueous solutions of pH 13 ± 0.1 and the carbon dosage was of 2 g L^{-1}.

The results obtained indicated very low Cu(I), Fe(II) and Fe(III) adsorption. From the D values defined as in Equation (4), the values of the separation factors Au(I)/M, defined as

$$\beta_{\frac{Au}{M}} = \frac{D_{Au}}{D_M}, \tag{5}$$

are summarized in Table 5 and show the excellent possibilities, β values greater than 1, of these nanomaterials to separate gold(I) from these others accompanying metals in the solution.

Table 5. Adsorption of different metal-cyano complexes and selectivity of the system.

Metal Pair	Complex	[M], mg g^{-1}	[M], mg L^{-1}	D, L g^{-1}	$\beta_{Au(I)/M}$
Au(I)	$Au(CN)_2^-$	2.8	4.4	0.64	64
Cu(I)	$Cu(CN)_4^{3-}$	0.03	3.2	0.01	
Au(I)	$Au(CN)_2^-$	3.1	3.9	0.79	7.2
Fe(II)	$Fe(CN)_6^{4-}$	0.25	2.3	0.11	
Au(I)	$Au(CN)_2^-$	3.3	3.5	0.94	7.2
Fe(III)	$Fe(CN)_6^{3-}$	0.29	2.2	0.13	

Time: 2 h.

3.3. Gold Adsorption from Chloride Media

The adsorption of Au(III) from HCl solutions by the multi-walled carbon nanotubes was also investigated. The adsorption equilibrium can be firstly described by the next general reaction:

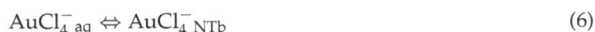

$$AuCl_4^-{}_{aq} \Leftrightarrow AuCl_4^-{}_{NTb} \tag{6}$$

where aq and Ntb represent the aqueous and solid phases, respectively. However, there is evidence of reduction of gold(III) to metallic gold, as the SEM image shows (Figure 2). As it is observed from this figure, dark particles appeared on the gold-loaded carbon nanotubes. These particles are of metallic gold as the EDS spectrum (Figure 3) showed two peaks at 9.7 keV (Lα) and 2.1 keV (Mα) characteristic of metallic gold. Very probably this reduction occurs on the carbon surface [17], being the reactions involved in the metal reduction:

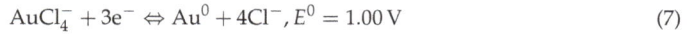

$$AuCl_4^- + 3e^- \Leftrightarrow Au^0 + 4Cl^-, E^0 = 1.00\,V \tag{7}$$

if the gold(III) reduces directly to metallic gold, or:

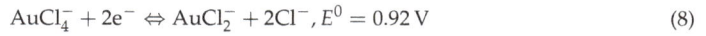

$$AuCl_4^- + 2e^- \Leftrightarrow AuCl_2^- + 2Cl^-, E^0 = 0.92\,V \tag{8}$$

$$AuCl_2^- + e^- \Leftrightarrow Au^0 + 2Cl^-, E^0 = 1.16\,V \tag{9}$$

if the reduction occurs via the formation of the $AuCl_2^-$ intermediate. In any case, the source for the electrons is the carbon nanotube:

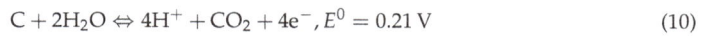

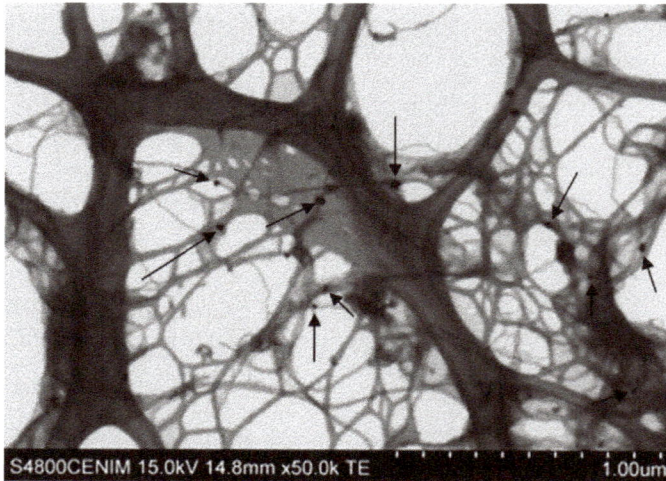

$$C + 2H_2O \Leftrightarrow 4H^+ + CO_2 + 4e^-, E^0 = 0.21\,V \tag{10}$$

Figure 2. SEM image of gold-loaded multi-walled carbon nanotubes. Arrows pointing to zero valent gold particles.

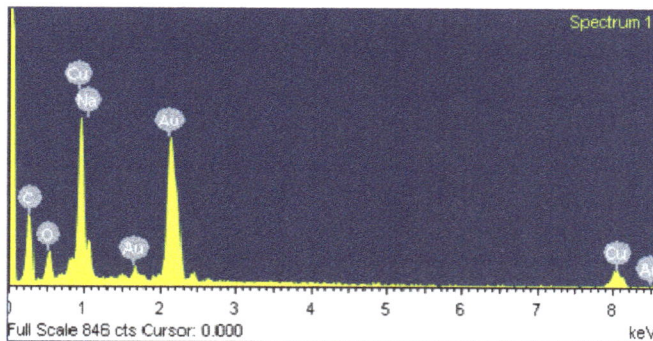

Figure 3. Energy Dispersive X-ray Spectroscopy spectrum of the dark particles (zero valent gold) of Figure 2.

3.3.1. Influence of the Stirring Speed in the Aqueous Phase

Experiments were performed to establish adequate hydrodynamic conditions. The adsorption of gold(III) was studied as a function of the stirring speed on the aqueous phase solution side and the results obtained are shown in Figure 4. Near constant adsorption for stirring speeds in the 500–750 min^{-1} range was obtained. Consequently, the thickness of the aqueous diffusion layer and the aqueous resistance to mass transfer were minimized. The diffusion contribution of the aqueous species to the mass transfer process was assumed to be constant. A stirring speed of 500 min^{-1} was maintained throughout all the investigation for the aqueous phase. The decrease of the percentage of gold(III) adsorption at stirring speeds above 750 min^{-1} is attributable to excessive speed resulting in local gold(III)-carbon nanotubes equilibrium, thus, decreasing the value of the percentage of the removal of this precious metal from the aqueous solution.

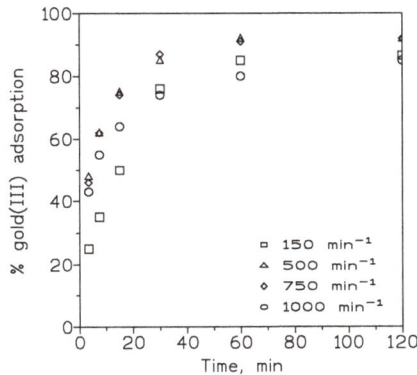

Figure 4. Influence of the stirring speed on gold(III) adsorption. Aqueous solution: 0.005 g L^{-1} Au(III) in 0.1 M HCl. Adsorbent dosage: 0.25 g L^{-1}. Time: 2 h.

3.3.2. Influence of Temperature

Experiments were performed to investigate the effect of this variable on gold(III) adsorption with an aqueous solution containing 0.005 g L^{-1} gold(III) at 0.1 M HCl and a carbon nanotubes dosage of 0.25 g L^{-1}. From the results obtained, it is shown that the increase of the temperature from 20 to 60 °C had no effect on gold adsorption, reaching 92 % after 30 min, in any case. However, the time to reach this maximum adsorption is temperature dependent (Figure 5).

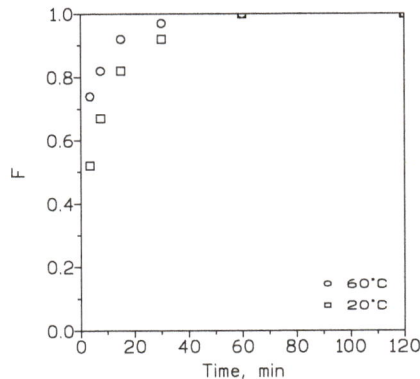

Figure 5. Influence of temperature on gold(III) adsorption. The F values were calculated as in Equation (3).

3.3.3. Influence of the Carbon Nanotubes Dosage

A series of experiments were performed using aqueous solutions containing 0.005 g L^{-1} gold(III) and 0.1 M HCl, whereas the carbon nanotubes dosage was varied from 0.05 to 0.5 g L^{-1}, in order to investigate the influence of this variation on gold(III) adsorption. Table 6 shows the variation of the percentage of gold adsorption and metal loading for different carbon nanotubes dosages. This shows that the increase of the amount of carbon nanotubes added to the solution increased the percentage of metal adsorbed onto the nanotubes.

Table 6. Influence of nanotubes dosage on gold(III) adsorption.

Dosage, g L^{-1}	% Gold Adsorption	Metal Uptake, mg g^{-1} [a]
0.5	99.5	10.0
0.25	92.0	18.4
0.13	51.0	19.6
0.05	31.0	25.8

[a] After 2 h.

3.3.4. Influence of the Initial Hydrochloric Acid Concentration

The variation in adsorption, as a function of the initial acid concentration, at 1 g L^{-1} carbon nanotubes dosage, has been studied when the aqueous phase contained 0.005 g L^{-1} gold at different HCl concentrations. The results are shown in Table 7, in which a near quantitative gold adsorption occurs at every initial HCl concentration and 2 h of contact between the aqueous solution and the adsorbent; though at shorter contact times, i.e., 30 min, an order in the sequence 0.1 M > 1 M> 10 M HCl was obtained and this is probably due to the existence of the HAuCl$_4$ species in the aqueous phase at higher HCl concentrations, against the presence of AuCl$_4^-$ species, at the more dilute hydrochloric acid solutions, which is more readily adsorbed than the former.

Table 7. Influence of initial HCl concentration on gold(III) adsorption.

HCl, M	% After 30 min	% After 2 h
0.1	99.5	99.5
1	94.0	99.5
10	66.5	99.5

3.3.5. Influence of the Adsorbent/Aqueous Phase Relationship on the Adsorption of Gold

The results concerning the adsorption of gold(III) from an aqueous phase containing 0.005 g L^{-1} Au(III) in 0.1 M HCl, at different volume of solution/carbon nanotubes weight ratios, revealed a change in the percentage of gold adsorption (92% at 1 h of contact) at the lowest solution/adsorbent relationship (4000) against the value of 51%, also at 1 h, obtained for a relationship of 8000. The gold uptakes were of 18.4 mg g^{-1} and 20.4 mg g^{-1} for the low and high ratios, respectively.

3.3.6. Separation of Gold (III) from Selected Base Metals

Since base metals are normally found in the gold-HCl bearing solutions, the selectivity of the present adsorption system against the presence of various metals in the aqueous phase was investigated by using an adsorbent dosage of 0.25 g L^{-1} and the aqueous phase containing 5.1 × 10^{-5} M Au(III), Fe(III), Cu(II) in 0.1 M HCl media. From the results obtained on the basis of binary solutions Au(III)-metal (Table 8), it is inferred that gold was preferably adsorbed over these base metals. The selectivity with respect to gold seems to be enough to separate this precious metal selectively from the base metals.

Table 8. Adsorption of gold(III) and base metals from acidic medium and separation factors.

Metal Pair	[a] Chloride Complexes	[M], mg g^{-1}	[M], mg L^{-1}	D, L g^{-1}	$\beta_{Au(III)/M}$
Au(III)	AuCl$_4^-$	28.8	2.8	10.3	103
Ni(II)	NiCl$^+$	0.23	2.9	0.1	
Au(III)	AuCl$_4^-$	24.8	3.8	6.5	81
Fe(III)	FeCl$_2^+$, FeCl^{2+}	0.22	2.7	0.08	
Au(III)	AuCl$_4^-$	27.6	2.9	9.5	119
Cu(II)	CuCl$^+$	0.26	3.1	0.08	

[a] At pH 1 and HCl 0.1 M. The values of D and β calculated as in Equations (4) and (5), respectively.

3.4. Adsorption Kinetics

To investigate the adsorption rate of both gold(I) and gold(III) onto the carbon nanotubes, the pseudo-first and pseudo-second order rate equations were used to fit the experimental data. The calculations concluded that the pseudo-second order model [18],

$$\frac{t}{[Au]_{r,t}} = \frac{1}{k_2 [Au]_{r,e}^2} + \frac{t}{[Au]_{r,e}}, \tag{11}$$

best fits to the experimental data both for the Au(I) and Au(III) systems. In this equation, k_2 is the rate constant, and $[Au]_{r,t}$ and $[Au]_{r,e}$ are the gold concentrations in the nanotubes at an elapsed time t and at equilibrium, respectively. Thus, the rate constants were estimated as 0.72 g mg^{-1} min^{-1} (r^2 = 1.000) for gold(I) and 0.014 g mg^{-1} min^{-1} (r^2 = 0.9997) for gold(III). These data suggest that chemical adsorption also contributes to physical adsorption to the overall metal adsorption onto the multi-walled carbon nanotubes [19].

3.5. Elution

After the adsorption of the metal onto the adsorbents, the next step should be the recovery of the metal from the adsorbent for a final recovery or dumping step. This desorption step, also called the elution step, is of equal importance as the adsorption step [20], though it is sometimes neglected by authors in their publications, i.e., References [2,7,12,21,22].

In the present investigation and in the case of gold(I), besides the methods described in the literature for gold desorption or elution from gold-loaded activated carbon [17], the elution of this element from the loaded carbon nanotubes can be accomplished using acidic thiourea solutions (1 g L^{-1} thiourea in 0.1 M HCl), reaching yields of near 65% of gold recovery (batch experiments at 20 °C and a volume of eluant/weight of gold-loaded carbon nanotubes relationship of 200) after 15 min of contact of the gold-loaded nanotubes and the eluant.

In the case of gold(III) and due to the presence of metallic gold in the nanotubes, the elution step may be considered as a dissolution problem rather than a true elution step. Thus in this case, the removal of gold from the nanotubes may be accomplished by the use of aqua regia, which dissolves the gold particles and renders a concentrated and pure gold solution from which gold can be recovered as gold nanoparticles (Figure 6), accordingly with the procedure described in the literature [23].

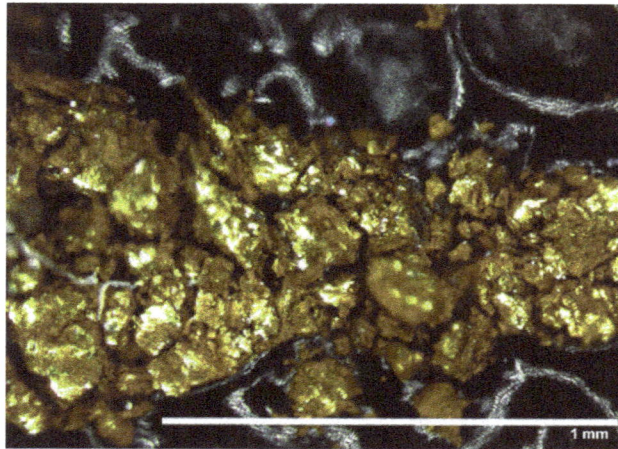

Figure 6. Zero valent gold nanoparticles obtained by sodium borohydride precipitation of the solutions resulted from the leaching with aqua regia of the gold-loaded multiwalled carbon nanotubes.

4. Conclusions

The experimental results indicate that it is possible to use multi-walled carbon nanotubes to remove $Au(CN)_2^-$ species at alkaline pH values. The adsorption of this anionic species is influenced by the aqueous pH values, enhancing the metal adsorption as the pH shifted to more acidic pH values and the aqueous solution/adsorbent relationship decreased its value. From the experimental data, the reaction enthalpy was estimated as -13 kJ mol^{-1}, indicating an exothermic adsorption reaction.

From acidic chloride media, this adsorbent can also be used to remove gold(III) at HCl solutions, retarding the adsorption as the initial aqueous acidity is increased. In this medium, the percentage of gold(III) adsorption is near quantitative after 1 h contact.

Moreover, a carbon-based nanotube technology has been developed for the recovery of gold from solutions obtained from the hydrometallurgical treatment of solid wastes and its separation from common and less valuable accompanying metals in the two aqueous media. Investigations in continuous mode (columns) will be further necessary to gain final knowledge of the performance of the carbon material (no loss of the adsorptive or elution properties) under several cycles of use.

Gold is finally rendered as zero valent gold nanoparticles.

Acknowledgments: The authors wish to thank the Agency CSIC (Spain) for support.

Conflicts of Interest: The auhor declare no conflicts of interest.

References

1. Iijima, S. Helical microtubules of graphitic carbon. *Nature* **1991**, *354*, 56–58. [CrossRef]
2. Hayati, B.; Maleki, A.; Najafi, F.; Gharibi, F.; McKay, G.; Gupta, V.K.; Harikaranahalli Puttaiah, S.; Marzban, N. Heavy metal adsorption using PAMAM/CNT nanocomposite from aqueous solution in batch and continuous fixed bed system. *Chem. Eng. J.* **2018**, *346*, 258–270. [CrossRef]
3. Desouky, A.M. Remove heavy metals from groundwater using carbon nanotubes grafted with amino compound. *Sep. Sci. Technol. (Philadelphia)* **2018**, *53*, 1698–1702. [CrossRef]
4. Sebastian, M.; Mathew, B. Multiwalled carbon nanotube based ion imprinted polymer as sensor and sorbent for environmental hazardous cobalt ion. *J. Macromol. Sci. Part A Pure Appl. Chem.* **2018**, *55*, 455–465. [CrossRef]

Appl. Sci. **2018**, *8*, 2264

5. Vilardi, G.; Mpouras, T.; Dermatas, D.; Verdone, N.; Polydera, A.; Di Palma, L. Nanomaterials application for heavy metals recovery from polluted water: The combination of nano zero-valent iron and carbon nanotubes. Competitive adsorption non-linear modeling. *Chemosphere* **2018**, *201*, 716–729. [CrossRef] [PubMed]

6. Żelechowska, K.; Sobota, D.; Cieślik, B.; Prześniak-Welenc, M.; Łapiński, M.; Biernat, J.F. Bis-phosphonated carbon nanotubes: One pot synthesis and their application as efficient adsorbent of mercury. *Fuller. Nanotub. Carbon Nanostruct.* **2018**, *26*, 269–277. [CrossRef]

7. Guo, X.; Feng, Y.; Ma, L.; Yu, J.; Jing, J.; Gao, D.; Khan, A.S.; Gong, H.; Zhang, Y. Uranyl ion adsorption studies on synthesized phosphoryl functionalised MWCNTs: A mechanistic approach. *J. Radioanal. Nucl. Chem.* **2018**, *316*, 397–409. [CrossRef]

8. Xu, J.; Cao, Z.; Zhang, Y.; Yuan, Z.; Lou, Z.; Xu, X.; Wang, X. A review of functionalized carbon nanotubes and graphene for heavy metal adsorption from water: Preparation, application, and mechanism. *Chemosphere* **2018**, *195*, 351–364. [CrossRef] [PubMed]

9. Oyetade, O.A.; Nyamori, V.O.; Jonnalagadda, S.B.; Martincigh, B.S. Removal of Cd^{2+} and Hg^{2+} from aqueous solutions by adsorption onto nitrogen-functionalized carbon nanotubes. *Desalin. Water Treat.* **2018**, *108*, 253–267. [CrossRef]

10. Burakov, A.E.; Galunin, E.V.; Burakova, I.V.; Kucherova, A.E.; Agarwal, S.; Tkachev, A.G.; Gupta, V.K. Adsorption of heavy metals on conventional and nanostructured materials for wastewater treatment purposes: A review. *Ecotoxicol. Environ. Saf.* **2018**, *148*, 702–712. [CrossRef] [PubMed]

11. Liu, D.; Deng, S.; Maimaiti, A.; Wang, B.; Huang, J.; Wang, Y.; Yu, G. As(III) and As(V) adsorption on nanocomposite of hydrated zirconium oxide coated carbon nanotubes. *J. Colloid Interface Sci.* **2018**, *511*, 277–284. [CrossRef] [PubMed]

12. Lu, F.; Astruc, D. Nanomaterials for removal of toxic elements from water. *Coord. Chem. Rev.* **2018**, *356*, 147–164. [CrossRef]

13. Pang, S.-K.; Yung, K.-C. Prerequisites for achieving gold adsorption by multiwalled carbon nanotubes in gold recovery. *Chem. Eng. Sci.* **2014**, *107*, 58–65. [CrossRef]

14. Alguacil, F.J.; Garcia-Diaz, I.; Lopez, F.; Rodriguez, O. Removal of Cr(VI) and Au(III) from aqueous streams by the use of carbon nanoadsorption technology. *Desalin. Water Treat.* **2017**, *63*, 351–356. [CrossRef]

15. Alguacil, F.J.; Lopez, F.A.; Rodriguez, O.; Martinez-Ramirez, S.; Garcia-Diaz, I. Sorption of indium(III) onto carbon nanotubes. *Ecotoxicol. Environ. Saf.* **2016**, *130*, 81–86. [CrossRef] [PubMed]

16. Alguacil, F.J.; Caravaca, C. Synergistic extraction of gold(I) cyanide with the primary amine Primene JMT and the phosphine oxide Cyanex 921. *Hydrometallurgy* **1996**, *42*, 197–208. [CrossRef]

17. Marsden, J.O.; House, C.I. Solution purification and concentration. In *The Chemistry of Gold Extraction*; SME Publication: Littleton, CO, USA, 2006.

18. Ho, Y.S.; McKay, G. Pseudo-second order model for sorption processes. *Proc. Biochem.* **1999**, *34*, 451–465. [CrossRef]

19. Chen, L.; Yu, S.; Liu, B.; Zuo, L. Removal of radiocobalt from aqueous solution by different sized carbon nanotubes. *J. Radioanal. Nucl. Chem.* **2012**, *292*, 785–791. [CrossRef]

20. Kupai, J.; Razali, M.; Buyuktiryaki, S.; Kecilic, R.; Szekely, G. Long-term stability and reusability of molecularly imprinted polymers. *Polym. Chem.* **2017**, *8*, 666–673. [CrossRef] [PubMed]

21. Rohanifar, A.; Rodriguez, L.B.; Devasurendra, A.M.; Alipourasiabi, N.; Anderson, J.L.; Kirchhoff, J.R. Solid-phase microextraction of heavy metals in natural water with a polypyrrole/carbon nanotube/1, 10–phenanthroline composite sorbent material. *Talanta* **2018**, *188*, 570–577. [CrossRef] [PubMed]

22. Azamat, J.; Hazizadeh, B. Removal of Cd(II) from water using carbon, boron nitride and silicon carbide nanotubes. *Membr. Water Treat.* **2018**, *9*, 63–68. [CrossRef]

23. Alguacil, F.J.; Adeva, P.; Alonso, M. Processing of residual gold(III) solutions via ion exchange. *Gold Bull.* **2005**, *38*, 9–13. [CrossRef]

applied
sciences

MDPI

Article

New Generation of Electrochemical Sensors Based on Multi-Walled Carbon Nanotubes

Thiago M. B. F. Oliveira [1] and Simone Morais [2,*

[1]   Centro de Ciência e Tecnologia, Universidade Federal do Cariri, Av. Tenente Raimundo Rocha,
      Cidade Universitária, Juazeiro do Norte 63048-080, CE, Brazil; thiago.mielle@ufca.edu.br
[2]   REQUIMTE–LAQV, Instituto Superior de Engenharia do Porto, Instituto Politécnico do Porto,
      Rua Dr. Bernardino de Almeida 431, 4249-015 Porto, Portugal
*   Correspondence: sbm@isep.ipp.pt; Tel.: +351-22-8340500; Fax: +351-22-8321159

Received: 13 August 2018; Accepted: 12 October 2018; Published: 15 October 2018

Abstract: Multi-walled carbon nanotubes (MWCNT) have provided unprecedented advances in the design of electrochemical sensors. They are composed by sp^2 carbon units oriented as multiple concentric tubes of rolled-up graphene, and present remarkable active surface area, chemical inertness, high strength, and low charge-transfer resistance in both aqueous and non-aqueous solutions. MWCNT are very versatile and have been boosting the development of a new generation of electrochemical sensors with application in medicine, pharmacology, food industry, forensic chemistry, and environmental fields. This work highlights the most important synthesis methods and relevant electrochemical properties of MWCNT for the construction of electrochemical sensors, and the numerous configurations and successful applications of these devices. Thousands of studies have been attesting to the exceptional electroanalytical performance of these devices, but there are still questions in MWCNT electrochemistry that deserve more investigation, aiming to provide new outlooks and advances in this field. Additionally, MWCNT-based sensors should be further explored for real industrial applications including for on-line quality control.

Keywords: nanomaterials; multi-walled carbon nanotubes; synthesis methods; electrochemical properties; electrochemical sensors; electroanalysis; sensing applications

---

1. Introduction

Several nanomaterials have been exploited for different purposes, but carbon nanotubes (CNT) and their composites emerged as the most exciting and versatile [1–5]. CNT are carbon allotropes with intermediate properties between graphite and fullerenes [6]. They are composed by sp^2 carbon units oriented as one (single-walled; SWCNT) or multiple concentric tubes (multi-walled; MWCNT) of rolled-up graphene, separated by ≈0.35 nm [4,6–8]. Both nanostructures had positive impacts for the applied sciences, but the latter has gained more attention due to the higher yield and lower production cost per unit, thermochemical stability, and ability to maintain or improve its electrical properties when submitted to different functionalization processes [9]. A general MWCNT representation and the possible π- and σ-covalent bonds established along the nanostructure are displayed in Figure 1. MWCNT also present remarkable active surface area, chemical inertness, high strength, and low charge-transfer resistance in both aqueous and non-aqueous solutions, making them suitable for numerous applications as nano-sized metallic and/or semiconducting components in transistors [10], energy storage [11], hybrid conductors [12], supercapacitors [13], thin-film coatings [2,5,6], field emitters [14], (bio)fuel cells [1–3,7], photovoltaic devices [1,3], electromagnetic shields [1,7], and electrochemical probes [1,2,5–7].

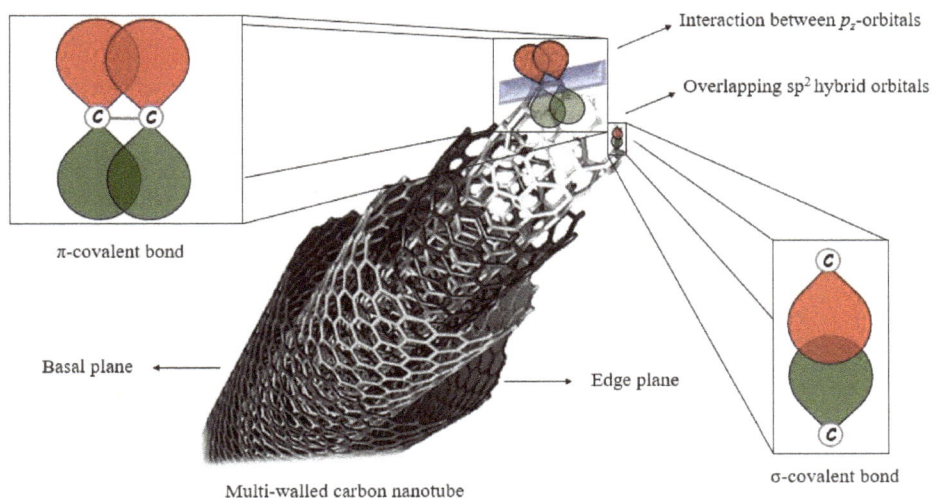

Figure 1. Representative structure of non-continuous multi-walled carbon nanotubes, emphasizing the different types of orbital interactions, as well as σ- and π-covalent bonds established along the 1D carbon networks.

MWCNT provide a series of functionalities to the electrochemical sensors, making them fantastic devices for analytical purposes. Electrochemical sensors are by definition, according the International Union of Pure and Applied Chemistry, "electrochemical devices that transform the effect of the electrochemical interaction analyte-electrode into a useful analytical signal" [15]; these effects may be promoted electrically or may be caused by a spontaneous interaction at the zero-current condition [15]. Thus, electrochemical sensors may be organized in four main subgroups: voltammetric, potentiometric, potentiometric solid electrolyte gas sensors, and chemically sensitized field effect transistor [15]. MWCNT associate high surface area (about 1600 m^2 g^{-1}) and exceptional electrical conductivity (current densities as high as 10^9 A cm^{-2}), which suit them to the miniaturization and portability of the systems [14].

Nowadays, electrochemical MWCNT-based sensors have demonstrated remarkable applications in medicine, clinical diagnostics, pharmacology, food industry, sanitary surveillance, occupational safety, forensic chemistry, and environmental analysis [1–3,6–9,15–18]. Other promising innovations can also be found in review papers that have been published over the last decade [6–9,15–20]. The assays are performed by various electrochemical techniques (voltammetry, amperometry, potentiometry, impedance spectroscopy, and their associations), which are selected according to the nature of the target-redox process [2,6,7,15–20]. MWCNT bring together high electronic conductivity, large specific surface area, chemical stability, biocompatibility, and ease of modification, improving the intensity and resolution at which the electroanalytical signals are observed on the working platforms [15–18]. However, the successful application requires a refined control of the sensor architecture and physicochemical properties, as well as the functionalization and/or surface modification of the carbon nanotubes [4,6–9]. Thus, in this study, these critical variables will be highlighted and discussed, addressing exciting issues from the MWCNT synthesis to their application in electrochemical biosensors design. With this goal in mind, the available literature concerning this subject in Thomson Reuters, *Web of Science* was reviewed from 2013 to March 2018.

2. MWCNT Synthesis Methods

Regarding synthesis methods, there are different paths that can be followed, such as electrical arch-discharge (EAD), laser ablation (LA), and numerous types of chemical vapor deposition (CVD), so that MWCNT conductor character varies according to the diameter and degree of helicity [4,7,21]. An overall layout of each system and its associated processes are shown in Figure 2. The yield rate of all aforementioned methods is more than 75% under appropriate operating conditions [22], but some questions about the structural quality, diameter uniformity, chirality, practicality, number of unitary operations, and production cost continue to divide opinions about the best strategy [4,21–24]. Researchers are continually trying to explore each possibility to optimize or innovate them for several new applications.

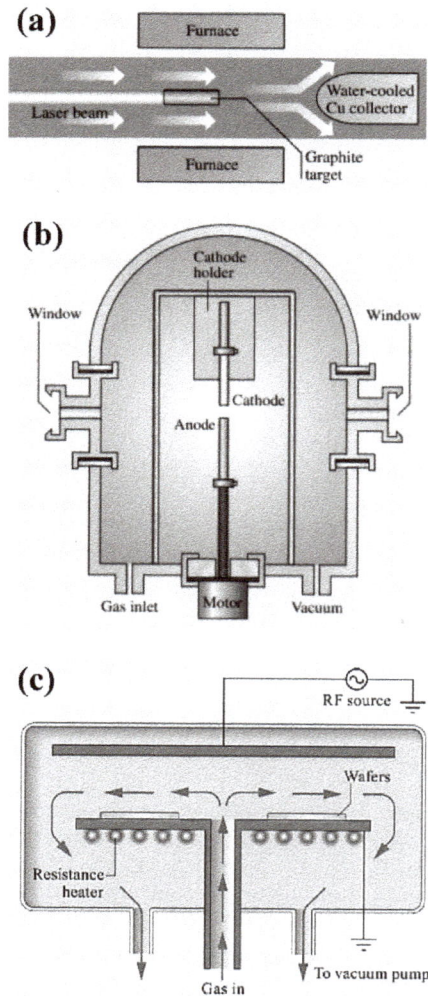

Figure 2. General layout of an (**a**) electrical arch-discharge (adapted by permission from Reference [23]), (**b**) laser ablation (adapted by permission from Reference [23]) and (**c**) chemical vapor deposition apparatus (adapted by permission from Reference [25]) used to synthesize multi-walled carbon nanotubes.

EAD is relatively less expensive, offers better yield quality, but involves high temperature for synthesis [21–23]. Triggering an electrical arc-discharge between two electrodes, a plasma composed by carbon and metallic catalysts vapor (iron, nickel, cobalt, yttrium, boron, cerium, among others) is formed under a rare gas atmosphere (helium or argon) [21–23]. The vapor content is a consequence of the energy transferred to the catalysts-doped anode, causing its erosion, while MWCNT are formed as a hard and consistent deposit on the cathode [23]. The LA method also represents a promising alternative to synthesize high-quality carbon nanotubes at room temperature, although it demands high instrumental and operational costs [21–23]. In this technique, a pulsed or continuous laser beam is focused on catalyst-based graphite pellet, which is placed at the center of a quartz tube filled with an inert gas and kept at 1200 °C. The radiant energy is enough to sublimate the carbon substrate, and this vapor is swept by the gas flow towards the conical water-cooled copper collector. MWCNT deposits are formed in the majority on the same collector, but appreciable amounts are also found on the quartz tube walls and graphite pellet backside. Both EAD and LA also produce in parallel other carbon phases and all products contain metallic impurities from the catalysts [23]. The small fractions of remaining metals do not compromise the performance of the electrochemical devices, and there are several purification processes proposed in the literature and by some commercial companies to remove undesirable phases [22–24]. Nevertheless, these procedures are based on oxidation under strong acidic conditions, which may significantly affect the integrity of the nanostructures.

The synthesis performed by CVD overcomes the aforementioned difficulties, guarantees large-scale production and, for these reasons, it has become more widely used to obtain MWCNT [22–25]. This technique involves either a heterogeneous (if a solid substrate is used) or a homogeneous process (supposing that reactions takes place in the gas phase) [23,25]. The main synthetic route explores the catalytic decomposition of a carbon-containing source on substrates containing transition metals and their composites (Au, Ag, Cu, Cr, Co, Fe, Mn, Mo, Ni, Pt, Pd, SiO$_2$, SiC, and ceramics), which enables the growth of aligned and dense arrays of nanotubes [23,24]. Alternative energy sources such as plasma and optical excitation can also be employed, giving the possibility to synthesize MWCNT at low-temperatures (450–1000 °C) compared to EAD and LA [23–25]. The CVD method also allows greater control of selectivity/type, homogeneity and size of the nanotubes (a few tens to hundreds of micrometers) in comparison to the others, although these characteristics are strongly dependent on the nature and structure of the catalysts, and the operational conditions of the synthesis. All these aspects affect the density and speed of electron transfer through the nanostructures that, in turn, directly impact the performance of the electrochemical sensors.

3. MWCNT Electrochemical Properties

MWCNT-based electrochemical sensors have remarkable advantages for analytical applications, such as enhanced detection sensitivity, low charge-transfer resistance (R_{tc}), electrocatalytic effect, and reduced fouling. It is also speculated that traces of metallic catalysts can add remarkable electroanalytical characteristics (positive synergistic effect) to the nanotubes, playing a supporting role for the development of the devices [6,7,15–26]. There are still controversies about the electron transfer mechanism along the structural network: Some researchers believe that the sidewalls of the CNT are inert (i.e., the edge–plane-graphite-like open ends and defect sites control the phenomenon) [26–28], but some studies conducted with two-dimensional SWCNT have shown evidence of charge-transfer on the basal-plane [7,23,25]. Although the uncertainties about the fundamental contributions of MWCNT in electrochemical sensors, the number of MWCNT successful applications in highly complex systems does not stop growing [2,9,17,26].

Understanding the size-dependent features is an indispensable requirement to explore the fascinating properties of nanotubes in analytical tools and to move the electroanalytical researches forward. In this sense, it must be assumed that the electronic band structure of CNT as a hexagonal lattice, where each carbon atom is covalently bonded to three others via sp^2 molecular orbitals (the fourth valence electron in a p_z-orbital hybridizes with all others to form a delocalized π-band) [27].

Like this, an even number of electrons are contained in the basic nanotube structure, whose conduction character will depend on its behavior as metal or semiconductor. This property can be assessed through the tight-binding electronic structure calculations, considering the translational vector (C_h) that connects two equivalent crystallographic sites (unitary vectors a_1 and a_2) and its coefficients (n_1 and n_2), according to the following Equation (1):

$$C_h = n_1 a_1 + n_2 a_2 \tag{1}$$

so that a metallic behavior is observed when the result is an integer multiple of three, while a non-metallic/semiconductor profile is observed for other cases; a direct consequence of the different conduction band gaps [6]. The electron-transfer along the MWCNT are also influenced by the tube diameter (d_t), chiral angle (θ), and basic translation vector (T), as shown in Figure 3. Thus, the manner in which a graphene sheet is rolled determines the main electronic properties of the resulting nanotube [6,15,23,27].

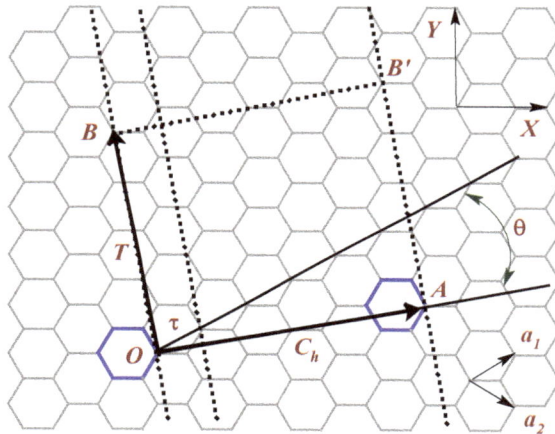

Figure 3. Honeycomb lattice structure of graphene and the main parameters that control the chirality and electronic properties of the resultant nanotube (adapted by permission from References [8,23]).

Observing the referred parameters on the tubulene lattice, d_t values can be determined by Equations (2) and (3), which relate this variable with the circumferential length of the nanotube (L) and a hexagonal network constant (a):

$$d_t = \frac{L}{\pi} = \frac{C_h}{\pi} = \frac{a\left(n_1^2 + n_1 n_2 + n_2^2\right)^{\frac{1}{2}}}{\pi} \tag{2}$$

considering $a_\circ = 1.42$ Å for carbon nanotubes [6]:

$$a = \sqrt{3}a_\circ = 2.49 \text{ Å} \tag{3}$$

Regarding the θ, the expression that describes its magnitude can be written as follows:

$$\cos\theta = \frac{2n_1 + n_2}{2\sqrt{n_1^2 + n_1 n_2 + n_2^2}} \tag{4}$$

The vector T is used to specify the orthogonal orientation to the C_h axis of the nanotube, and its value can be calculated by:

$$T = \frac{(2n_2 + n_1)a_1 - (2n_1 + n_2)a_2}{d_k} \tag{5}$$

where d_k ranging from d (when $n_1 - n_2$ and d_k is not a multiple of $3d$) to $3d$ (when $n_1 = n_2$ and d_k is a multiple of $3d$) for chiral tubulenes, considering d the greatest common divisor for the coefficients n_1 and n_2 [6,15,23,27].

In addition, a translation vector (τ) is also necessary to delimit the unitary cell of the nanotube. It can be represented by Equations (6)–(8),

$$\tau = t_1 a_1 + t_2 a_2 \tag{6}$$

$$t_1 = \frac{(2n_2 + n_1)}{d_r} \tag{7}$$

$$t_2 = \frac{-(2n_1 + n_2)}{d_r} \tag{8}$$

knowing that d_r (d–$3d$) is the greatest common divisor of the respective coefficients. In summary, the limits of the 1D nanotube unit cell (plane segment $OAB'BO$) are demarcated by the vectors C_h and T, whereas a_1 and a_2 define the area of the 2D graphene unit cell [6,27]. This sequence of equations is very important for the development of electrochemical sensors because it allows to determine the number of electric and phononic bands in the carbon nanostructures, i.e., the magnitude of the electrical conduction through the nanomaterial.

The influence of the band structure on the nanotube conduction bands is defined in the context of the Fermi velocity (v_F).

$$v_F = \frac{\hbar k_F}{m} \tag{9}$$

where $\hbar$ is the reduced Planck constant, k_F is the Fermi wavevector, and m represents the mass of the nanostructure under study. These values are not well-defined in cases where the Fermi surface is non-spherical, but it is estimated to be $v_F \sim 10^6$ ms^{-1} for semiconductors or even higher for conductors [27]. This prevision corroborates the observation recorder in the density of states diagram for SWCNT with different chirality, since at the Fermi energy level (E_F) there is a finite electronic transition for metallic structures (first van Hove optical transition), but a zero-band gap for semiconductors (second van Hove optical transition). A general description of the band structures and density of states for metallic and semiconductor nanotube is presented in Figure 4. Thus, if we imagine that MWCNT are composed by several coaxial SWCNT, it might be expected that they are not strictly a 1D-type conductor [7,27]. Likewise, when the nanotubes have different chirality, the band structure undergoes alterations, and the resultant coupling between the electronic states at the E_F and phonon modes may cause superconductivity [4,7,27]. Despite the similarity with metallic materials, the carrier density in carbon nanotubes (1D quantum wires) is much lower and novel physical electron-electron phenomena (such as spin-charge separation and suppression of the electron tunneling density of states) should be considered to justify the possible discrepancies, as predicted by Luttinger liquid theory [23,26–28].

Some electrochemical studies showed a lower R_{tc} for MWCNT containing structural defects and/or traces of metallic impurities (catalyst nanoparticles), so that the current flow occurs mainly through the outer most nanotube cylinder [26–28]. Only one of the concentric tubes needs to exhibit this behavior for the overall electronic properties to be essentially metallic-low quantum capacitance values [7,27]. It is believed that if d_t is smaller than the elastic mean free path, the one-dimensional ballistic transport predominates, but the opposite characterizes a two-dimensional diffusive current flow [7,26–28]. In addition, nanotubes with larger d_t have a greater density of structural defects, besides facilitating their modification and functionalization, which are interesting requirements for the development of electroanalytical devices [2,6–9]. The mean length of the nanotubes also has a significant effect on the electron-transfer rate; they are inversely proportional properties [22,24,27]. Vertically aligned structures also exhibit superior performance compared to those randomly immobilized [2,16,25–28].

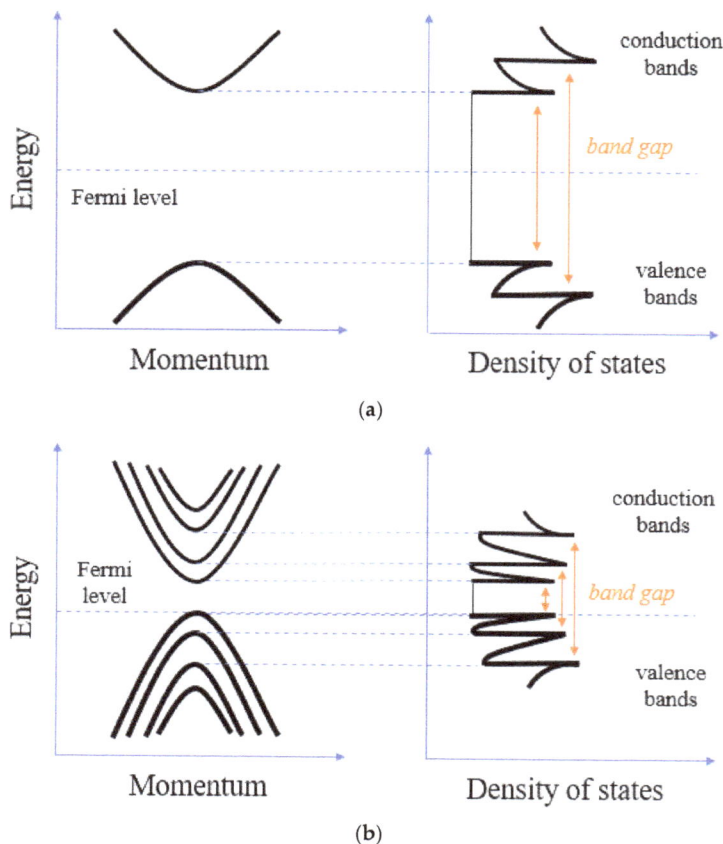

Figure 4. Idealized models of band structures and density of states for (**a**) metallic and (**b**) semiconducting nanotubes.

Therefore, MWCNT provide special characteristics in electrochemical sensors both in aqueous and non-aqueous electrolytes (electrocatalytic properties, high and fast electron-transfer, signal amplification and stability, among other particularities observed in different prototypes) when compared with the conventional carbonaceous electrodes (such as glassy carbon, highly-oriented pyrolytic graphite, carbon microfiber, and boron-doped diamond) [26–28]. However, the true reasons and fundamental electrochemical properties responsible for the reported performances (both when nanotubes are used alone or in (nano)composites) are still not fully elucidated and demand further research [2,4,6,9,17].

4. Overview of MWCNT Applications in Electrochemical Sensors

Different and exciting properties are obtained with macro-, micro-, and nano-structured carbon derivatives [4,17], but the implementation of the latter in the configuration of electrochemical sensors brought new opportunities to detect trace concentrations (μmol L^{-1} to pmol L^{-1}) of numerous analytes (active principle and metabolites of drugs, personal care products, flame retardants, gases, surfactants, industrial additives, polycyclic aromatic hydrocarbons (PAHs), polychlorinated biphenyls (PCBs), hormones, pesticides, among others) in complex matrices (natural waters, soils, foodstuffs, biological fluids, fuels, wastes, and even in living organisms) [1–17]. This trend led to the development of the newest category of electroanalytical devices (third generation/way), i.e., nanotools for nanoscale analysis [2,16]. The basic mechanism for the operation of these tools is based on the conversion of redox processes, registered at the sensor/electrolyte interface, onto quantitative or semi-quantitative information. MWCNT had an important role to reach such advances, and this fact can be easily attested by the exceptional growth of scientific reports in this field [2–6,8,16,17]. MWCNT's large active surface area, facile electron-transfer with a variety of molecules, and their use in nano-composites (mixed with conductive polymers, nanoparticles and active biomolecules), as well as their high potential for device miniaturization, have been enhancing the sensitivity of current methodologies and providing new sensing opportunities [2,17,19–22]. The multichannel electron ballistic transport and the presence of holes/defects along the structures, especially when functionalized with carboxyl groups and their derivatives, further enhance the previously mentioned properties [2,4–9]. When converting a graphene sheet to a nanotube, one electron of the carbon atoms remains in *p*-orbitals, organized between valence (π) and conduction (π^*) bands with very low bandgap (a partial σ–π hybridization), which favors a fast electron transfer. The density of states in MWCNT structure depends on diameter ($\approx$2–100 nm), chirality/configuration (concentric, herringbone, and bamboo) and nature (semiconductor, conductor, and superconductor) [1,7,15]. In addition, the strength of C−C bonds is among the most robust and, when they are organized in the nanotube form, these interactions also provide flexibility and resistance to rupture, increasing the versatility and number of possible applications of these nanomaterials [4,7,16].

Electrochemical sensing of a wide variety of chemically, biologically, and environmentally important analytes was investigated on MWCNT (chemical purity ≥95%)—based solid electrodes, as shown in Table 1. Numerous traditional and modern pharmaceuticals, for instance, are electroactive on these sensors and relevant contributions have been reported, either in stationary or in flow-through systems. Mao et al. [29] fabricated an electrochemical sensor for metronidazole (antibacterial and anti-inflammatory compound employed for the treatment of protozoal diseases) based on MWCNT and chitosan (CTS)-nickel complex modified glassy carbon electrode (GCE) by the self-assembly technique. The polymer matrix facilitates the immobilization of the nanocomposite on the electrode support, while the association between nickel and carbon nanomaterials increases the current intensity of the electroanalytical signal. Using differential pulse voltammetry (DPV), Ni-CTS/MWCNT/GCE also showed electrocatalytic effect toward metronidazole reduction, making the method highly selective to its analysis in pharmaceutical and biological samples, with a low detection limit (LOD; 0.025 μmol L^{-1}) and suitable reproducibility and stability. Holanda et al. [30] reported a simple methodology to quantify acetaminophen (a compound with analgesic and antipyretic action) in commercial formulations, using a GCE modified with gold nanoparticles (Au-NP), functionalized MWCNT (*f*-MWCNT) and cobalt phthalocyanine (Co-Pht) as working sensor. Using *f*-MWCNT, a lower system capacitance and better stability of the sensor was observed. The quantification of acetaminophen on Co-Pht/*f*-MWCNT/Au-NP/GCE was carried out by square-wave voltammetry (SWV) that, in turn, showed a sensitivity (LOD = 0.135 μmol L^{-1}) and reproducibility similar to the standard ultraviolet-visible spectrophotometric procedure (at the wavelength of 257 nm). Montes et al. [31] studied the superior electrochemical activity of MWCNTs with shorter dimensions (diameter $\times$ length: 100–170 nm $\times$ 5–9 μm and 6–9 nm $\times$ 5 μm) towards the oxidation of propionic acid derivatives, which are drugs with analgesic, antipyretic, and anti-inflammatory properties. Electrochemical impedance spectroscopy (EIS) and cyclic voltammetric (CV) data showed better electrocatalysis effect, R_{tc} and signal amplification for the sensor constructed with the shorter diameter nanostructures due to the increased defect density revealed by Raman spectroscopic measurements. The developed sensor (MWCNT (diameter $\times$ length: 6–9 nm $\times$ 5 μm)/GCE) was adapted to a flow-injection analysis system (FIA) and provided a sensitivity and accuracy to determine ibuprofen (LOD = 1.9 μmol L^{-1})

by amperometry comparable to that obtained by capillary electrophoresis. Pavinatto et al. [32] used a fluorine doped tin oxide (FTO) electrode, coated with nanostructured layer-by-layer films of CTS@MWCNT (Figure 5), as a sensor for 17-α-ethinylestradiol, that is a synthetic estrogen widely used as an oral contraceptive. The electron-transfer on the sensor reduced dramatically in the presence of the nanotubes, improving its electroanalytical performance and enabling assays at very low concentrations. Using the irreversible and adsorption-controlled oxidation process of the estrogen onto the three-bilayer CTS@MWCNT/FTO, it was possible to determine it by SWV (LOD = 0.09 μmol L^{-1}) without significant effect of the most common interfering compounds present in human urine (urea, uric, and ascorbic acids, glucose, NaCl, KCl, and NH$_4$Cl). Chen et al. [33] used a nanocomposite based on MWCNT and 1-octyl-3-methyl-imidazolium hexa-fluorophosphate ionic liquid (IL) to modify a GCE, to develop a chiral electrochemical sensor for the enantiomeric recognition of propranolol, using the linear sweep voltammetry (LSV) technique. The authors did not present a LOD but declared the success of the proposed method to evaluate the enantiomeric purity of reagents and wastewater treatment efficiency. Other promising MWCNT-based electrochemical devices were also reported to successfully analyze hydrochlorothiazide and triamterene [34], mangiferin and icariin [35], natamycin [36], omeprazole [37], gentamicin [38], and many other drugs [2,6,9,17–20], attesting their electroanalytical performance and effectiveness.

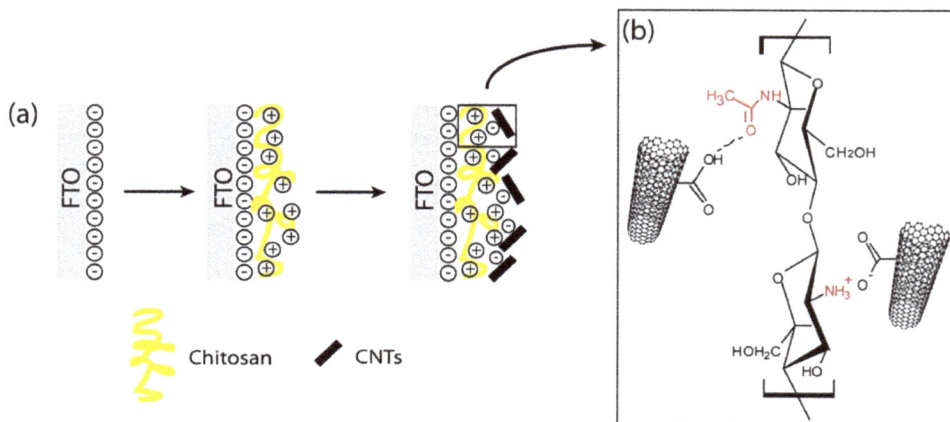

Figure 5. (**a**) Schematic representation of the layer-by-layer and (**b**) supramolecular interaction between chitosan and multi-walled carbon nanotubes (MWCNT) for 1-bilayer film (adapted by permission from Reference [32]).

The growing trend of using miniaturized systems for the electroanalysis of bioactive molecules in living organisms through less invasive tests has positive repercussion and wide acceptance [2,17]. From this perspective, glucose sensors have been receiving special attention, since these analytic tools can be used as an effective way to diagnose diabetes, hypoglycemia, and other metabolic diseases in real time, using human blood and other secreted fluids as study matrix. Başkaya et al. [39] fabricated a non-enzymatic glucose sensor using a hybrid film (composed by *f*-MWCNT and highly monodisperse nickel nanoparticles) immobilized on GCE, which was able to determine this target in human blood serum samples by amperometry, with sensitivity (LOD = 0.021 μmol L^{-1}) and stability similar to one of commercial enzymatic biosensors. The results also support the positive synergistic effects that are observed when different nanomaterials are associated into the same matrix, enhancing the capacity and versatility of the electrochemical sensors. Purine bases are also crucial biomolecules to the immune function and abnormal changes causes serious disorders, such as cancer, epilepsy, lupus erythematous, renal calculi, and cardiovascular alterations. Wang et al. [40] prepared a nanocomposite by mixing

MWCNT and copper-nickel hybrid nanoparticles (Cu@Ni-NP), which was deposited onto a GCE surface, with the goal of monitoring adenine and guanine in ds-DNA from mice brain tissues. Using DPV under the optimized conditions, the Cu@Ni-NP/MWCNT/GCE sensor achieved a LOD of 0.17 µmol L^{-1} and 0.33 µmol L^{-1} for guanine and adenine, respectively, attesting to its potential to quantify purine bases in complex samples. Likewise, sensors for point-of-care detection of DNA damage in medical diagnostics are also on the rise. Li and Lee [41] investigated the performance of *f*-MWCNT (functionalization with oxygen-containing groups to improve the immobilization and stability of the bioelement) as an additive in electrochemical DNA sequence differentiation ability. The electroanalytical signal achieved by DPV was highly selective and modulated between the probe (5'-GTG TTG TCT CCT AGG TTG GCT CTG-3'; 24-base fraction of the p53 gene, used as a marker for breast cancer) and its complementary sequence (5'-CAG AGC CAA CCT AGG AGA CAA CAC-3'). The detection capacity (LOD = 141.2 pmol L^{-1}) was more than two orders better than the previously reported for graphene functionalized (*f*-Grf) based sensors, even in the presence of single (5'-CAG AGC CAA CCT CGG AGA CAA CAC-3') and all-base mismatched pairs (5'-ATA TCG ACC TTG GCC GAG ACG GTG-3'). Great effort has also been directed towards the development of electrochemical devices capable of early detection of immune deficiency diseases, such as acquired immune deficiency syndrome (AIDS) caused by the HIV-retrovirus. Ma et al. [42] proposed a device composed by a GCE modified with MWCNT and molecularly imprinted polymers (MIPs) based on CTS for the determination of HIV-p24 protein in real human serum samples (Figure 6). Using DPV and the developed sensor (HIV-p24/CTS/MWCNT/GCE), a low limit of detection (0.083 pg cm^{-3}), good selectivity, repeatability, reproducibility, stability, and accuracy were reached for the proposed electroanalysis; validation was performed by comparative studies with enzyme-linked immunosorbent assay (ELISA). The outstanding ability of MWCNT-based electrochemical sensors to simulate redox reactions that occur in vivo has also been demonstrated with the electroanalysis of hemoglobin [43], amino acids [44], cholesterol [45], bilirubin [46], neurotransmitters, and related compounds [47–49], and even microorganisms [50], in mammalian fluid samples, achieving great results and contributing to the early diagnosis and treatment of several related diseases.

The occurrence of pesticides in foodstuffs, natural water, soils, among other matrices of environmental relevance, constitute a worldwide concern. Sipa et al. [51] associated square-wave adsorptive stripping voltammetry (SWAdSV) and a modified sensor (β-cyclodextrins/MWCNT/GCE) to determine trace concentrations of the pesticide dichlorophen. The excellent electroanalytical performance of the system, mainly in the presence of nanotubes (LOD = 4.4 × 10^{-8} mol L^{-1}), allowed to monitor this vermicide in river water (Pilica River, in Poland) with good selectivity and absence of natural and anthropogenic interfering compounds. Özcan and Gürbüz [52] developed a simple and sensitive sensor for the voltammetric determination of the herbicide clopyralid in urine, river water, sugar beet, wheat, and herbicide formulations (Phaeton® and Lontrel™). This novel sensor was built by modification of a GCE with a nanocomposite containing acid-activated MWCNT (oxygen-functionalized nanostructures with very low R_{tc}) and fumed silica, which, when associated with DPV, attained a LOD of 0.8 nmol L^{-1}, confirming the sensitivity of the proposed method. Ghodsi and Rafati [53] reported a strategy to develop a voltammetric sensor for diazinon (an organophosphate insecticide) based on TiO$_2$ nanoparticles (TiO$_2$-NP) and MWCNT nanocomposite assembled onto a GCE. Under optimum conditions, TiO$_2$-NP@MWCNT/GCE reached a LOD of 3.0 nmol L^{-1}, using both CV and SWV, and this performance was explored for the electroanalysis of the pesticide in well and tap water. Other mentioned advantages were the fast response time, good repeatability, stability, besides the fast and inexpensive electrode modification. Sensitive electrochemical sensors that associate the unique properties of MWCNT and metallic nanoparticles were also proposed by Wei et al. [54] and Ertan et al. [55], aiming to electroanalysize trichlorfon insecticide (LOD = 4.0 × 10^{-7} mol L^{-1}, using Nafion®/TiO$_2$-NP@MWCNT@carboxymethyl chitosan/GCE) and simazine herbicide (LOD = 2.0 × 10^{-11} mol L^{-1}, using MIP/platinum nanoparticles@polyoxometalate@*f*-MWCNT/GCE), respectively, reinforcing the positive synergistic effect between these nanostructures for the improvement of the electroanalytical methodologies.

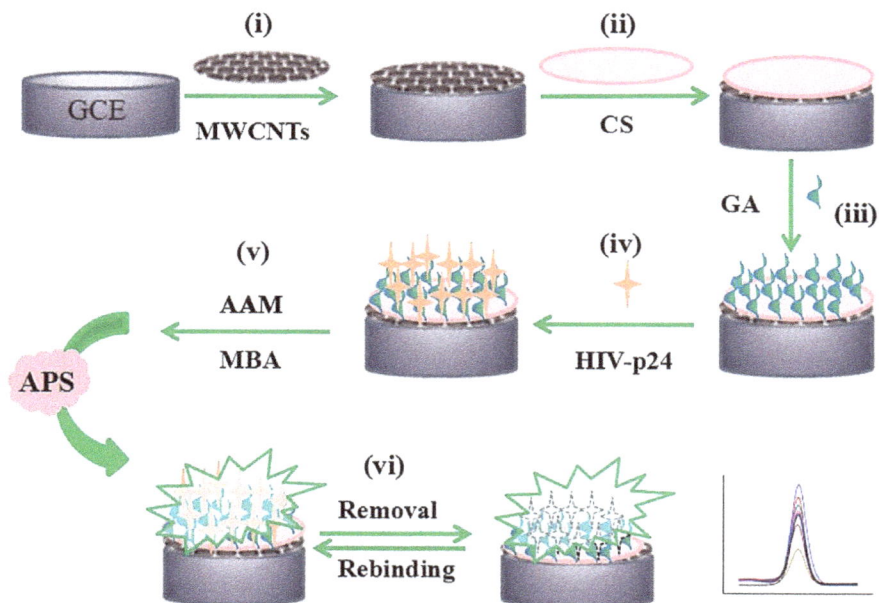

Figure 6. Detailed procedure diagram for fabrication of the HIV-p24/CTS/MWCNT/GCE sensor (adapted by permission from Reference [42]).

The wide active area and low R_{tc} observed for MWCNT, and related (nano)composites, have been providing important advances on the single and simultaneous electroanalytical quantification of metallic cations (Cd^{2+} [56], Pb^{2+} [56], Mn^{2+} [57], and Na^+ [58]) and anions (SO_3^{2-} [59], NO_2^- [60], and BrO_3^- [60]) in varied and complex matrices (wastewater [57], drinking water [56], and groundwater [58,59]). Trace concentrations of persistent organic pollutants (sunset yellow [61], tartrazine [62], and luteolin [62] dyes), as far as gases (chlorine [63], carbon dioxide [64], and methanol vapor [65]) and industrial by-products (glycerol [66], bisphenol A [67], hydrazine [68], and hydrogen peroxide [69]) are now being monitored with MWCNT-based sensors in a quick, reproducible, reliable, and cost-effective way when compared to the traditional analytical protocols, including those based on electrochemical devices from previous generations.

Table 1. Configuration, analytical performance and application of electrochemical sensors based on multi-walled carbon nanotubes.

Sensor	Modification Procedure	Analyte(s)	Technique(s)/Detection Limit	Application	Stability	Reference
Pharmaceuticals						
Ni-CTS/MWCNT/GCE	GCE modified with MWCNT and Ni-CTS complex through drop coating and self-assembly, respectively	metronidazole	DPV/0.025 μmol L^{-1}	tablet and biological samples	81% after one month	[29]
Co-Pht@f-MWCNT/Au-NP/GCE	suspension of Co-Pht and f-MWCNT immobilized on Au-NP modified GCE by drop coating	acetaminophen	SWV/0.135 μmol L^{-1}	commercial formulations	n.r.	[30]
MWCNT(shorter diameter)/GCE	MWCNT (diameter × length: 6–9 nm × 5 μm) dropped on GCE	ibuprofen	CV/1.90 μmol L^{-1}	tablet and liquid commercial formulations	n.r.	[31]
Three bilayer MWCNT@CTS/FTO	FTO coated with nanostructured Layer-by-Layer films of MWCNT@CTS	17-α-ethinylestradiol	SWV/0.09 μmol L^{-1}	synthetic urine samples	n.r.	[32]
MWCNT@IL/GCE	immobilization of MWCNT and IL (1-octyl-3-methyl-imidazolium hexa-fluorophosphate) nanocomposite on GCE	propranolol	LSV/n.r.	commercial reagent and wastewater	n.r.	[33]
MWCNT/GCE	suspension of MWCNT dropped on GCE	hydrochlorothiazide and triamterene	ASV/2.8 × 10^{-8} and 2.9 × 10^{-8} mol L^{-1} for hydrochlorothiazide and triamterene, respectively	hemodialysis samples	n.r.	[34]
Au@Ag-NP/MWCNT-SCSs/GCE	layer-by-layer assembly of Au@Ag-NP and MWCNT-SCSs on GCE	mangiferin and icariin	DPV/0.017 μmol L^{-1} for both compounds	*Rhizoma anemarrhenae*, *Artemisia capillaris Herba* and *Epimedium macanthum* samples	≤95.1% after one month	[35]
3D-Grf@MWCNT/GCE	electrodeposition of 3D-Grf@MWCNT suspension on GCE	natamycin	LSV/1.0 × 10^{-8} mol L^{-1}	red wine and beverage samples	94.6% after two weeks	[36]
Fe$_3$O$_4$-NP@MWCNT/PDDA/GCE	casting of PDDA modified GCE with Fe$_3$O$_4$-NP@MWCNT hybrid film	omeprazole	LSV/15 nmol L^{-1}	tablet, capsules, wastewater, serum, and urine	92.1% after three weeks	[37]
Calixarene/MWCNT/SPE	dip coating of graphite-based SPE in composite matrix of calixarene and MWCNT	gentamicin sulphate	potentiometry/7.5 × 10^{-8} mol L^{-1}	dosage forms and spiked surface water samples	n.r.	[38]
Biologically active molecules						
Ni-NP@f-MWCNTs/GCE	drop coating of hybrid film (Ni-NP and f-MWCNT) on GCE	glucose	CV and amperometry/0.021 μmol L^{-1}	human blood serum samples	practically constant signal after 1000th cycle	[39]
Cu@Ni-NP/MWCNT/GCE	immobilization of hybrid film of Cu@Ni-NP and MWCNT on GCE	guanine and adenine	DPV/0.17 μmol L^{-1} and 0.33 μmol L^{-1} for guanine and adenine, respectively	ds-DNA from mice brain tissues	96.7% for 30 days	[40]
Gold electrode	measurements with unmodified gold electrode, keeping f-MWCNT additive and DNA sequence in electrolyte solution	breast cancer marker (5'-GTG TTG TCT CCT AGG TTG CGT CTG-3'; 24-base fraction of the p53 gene)	DPV/141.2 pmol L^{-1}	solution containing the complementary sequence (5'-CAG AGC CAA CCT AGG AGA CAA CAC-3')	n.r.	[41]

Table 1. *Cont.*

Sensor	Modification Procedure	Analyte(s)	Technique(s)/Detection Limit	Application	Stability	Reference
HIV-p24/MIP/MWCNT/GCE	HIV-p24 crosslinking MIP (acrylamide functional monomer, N,N'-methylenebisacrylamide as crosslinking agent and ammonium persulphate as initiator) immobilized on MWCNT/GCE	HIV-p24 protein	DPV/0.083 pg cm^{-3}	real human serum samples	98.6% after 10 days	[42]
MIP/MWCNT/Cu	MIP (itaconic acid monomer, ethylene glycol dimethacrylate cross-linker and α,α'-azobisisobutyronitrile as initiator) on MWCNT modified Cu electrode	hemoglobin	potentiometry/1.0 µg mL^{-1}	human bile juice and urine samples	6 months without significant change in the electrode performance	[43]
Cu-MP@polyethylenimine/MWCNT/GCE	Cu-MP dispersed in polyethylenimine and dropped on MWCNT/GCE	amino acids, albumin and glucose	SWV and amperometry/0.10–0.37 µmol L^{-1} for the amino acids (L-cystine, L-histidine and L-serine); 1.2 mg mL^{-1} for albumin; and 182 nmol L^{-1} for glucose	pharmaceutical products and beverages	7.7% RSD after 10 successive calibration plots using the same surface	[44]
MIP/Au-NP/MWCNT/GCE	Au-NP electrodeposited on MWCNT/GCE, and assembled with MIP (tetrabutylammonium perchlorate)	cholesterol	DPV/3.3 × 10^{-14} mol L^{-1}	n.r.	91.7% after one month	[45]
MWCNT/SPE	MWCNT films casted onto SPE	bilirubin	CV/9.4 µmol L^{-1}	n.r.	n.r.	[46]
Grf-Ox@MWCNT/GCE	drop coating of Grf-Ox@MWCNT suspension on GCE	catechol and dopamine	CV/n.r.	n.r.	n.r.	[47]
Phenazine methosulfate/3-aminophenylboronic acid/f-MWCNT/GCE	drop coating of phenazine methosulfate and 3-aminophenyl boronic acid on f-MWCNT/GCE	NADH	amperometry/0.16 µmol L^{-1}	human serum	96.7% after five consecutive measurements	[48]
HPU/β-CD/MWCNT@Nafion®/GCE	layer-by-layer of HPU, β-CD and composite film (MWCNT@Nafion®) on GCE	uric acid	amperometry/n.r.	n.r.	n.r.	[49]
Microorganisms						
MWCNT@Nafion®/GCE	dip coating of composite suspension (MWCNT@Nafion®) on GCE	Enterotoxigenic *Escherichia coli* F4 (K88)	SWV/6 × 10^4 CFU mL^{-1}	swine stool samples	n.r.	[50]
Pesticides						
β-CD/MWCNT/GCE	β-CD and MWCNT composite suspension dropped on GCE	dichlofenthion	SWAdSV/4.4 × 10^{-8} mol L^{-1}	river water	93.9% after one week	[51]
Fumed silica/acid-activated MWCNT/GCE	drop coating of a nanocomposite suspension (Fumed silica and acid-activated MWCNT) on GCE	clopyralid	DPV/0.8 nmol L^{-1}	urine, river water, sugar beet, wheat, and herbicide formulations (*Phacton* and *Lontrel*)	91% after three weeks	[52]
TiO₂-NP@MWCNT/GCE	TiO₂-NP@MWCNT nanocomposite dropped on GCE	diazinon	CV an SWV/3.0 nmol L^{-1}	well and tap water	89% after 28 days	[53]
Nafion®/TiO₂-NP@MWCNT @carboxymethyl chitosan/GCE	Nafion® assembled on composite film (TiO₂-NP@MWCNT @carboxymethyl chitosan) previously immobilized on GCE	trichlorfon	DPV/4.0 × 10^{-7} mol L^{-1}	apple, mushroom, and cucumber	98% after one week	[54]
MIP/Pt-NP@polyoxometalate@f-MWCNT/GCE	MIP (pyrrole in the presence of the analyte) assembled on hybrid film (Pt-NP@polyoxometalate@f-MWCNT) immobilized on GCE	simazine	DPV/2.0 × 10^{-11} mol L^{-1}	wastewater samples	96.9% after 45 days	[55]

Table 1. *Cont.*

Sensor	Modification Procedure	Analyte(s)	Technique(s)/Detection Limit	Application	Stability	Reference
Metallic cations						
BiF/Grf-Red/MWCNT/SPE	layer-by-layer of BiF, Grf-Red and MWCNT on SPE (gold support)	Cd^{2+} and Pb^{2+}	SWV/0.6 ppb for Cd and 0.2 ppb for Pb	drinking water	n.r.	[56]
Mn²⁺-imprinted polymer /IL@CTS@MWCNT/GCE	thermal immobilization of Mn^{2+}-imprinted polymer on composite layer (IL@CTS@MWCNT) dropped on GCE	Mn^{2+}	SWAdSV/0.15 μmol L^{-1}	wastewater	94.8% after two weeks	[57]
β-NiOₓ/MWCNT-modified CPE	electrodeposition of hybrid film of β-NiOₓ on MWCNT-modified CPE	Na^{+}	SWV/9.86 × 10^{-8} mol L^{-1}	groundwater	practically constant signal for more than five hundred consecutive cycles	[58]
Anions						
f-MWCNT/GCE	drop coating of f-MWCNT (COOH-functionalized structures) as suspension on GCE	SO_3^{2-} and NO_2^{-}	DPV/215 nmol L^{-1} for SO_3^{2-} and 565 nmol L^{-1} for NO_2^{-}	groundwater	≥96.4% after one week	[59]
Ag-NP@MWCNT/GCE	drop coating of nanocomposite (Ag-NP@MWCNT) on GCE	BrO_3^{-}	amperometry/n.r.	n.r.	n.r.	[60]
Dyes						
Grf-Ox@MWCNT/GCE	suspension of Grf-Ox@MWCNT immobilized on GCE by drop coating	sunset yellow and tartrazine	LSV/0.025 μmol L^{-1} for sunset yellow and 0.010 μmol L^{-1} for tartrazine	orange juice	89–93% after one month	[61]
Poly(crystal violet)/MWCNT/GCE	electropolymerization of crystal violet on MWCNT/GCE	luteolin	DPV/5.0 × 10^{-9} mol L^{-1}	Chrysanthemum samples	93% after one month	[62]
Gas/Vapor						
Hexa-decafluorinated zinc phthalocyanine @f-MWCNT/SPE (gold support)	drop coating of the composite (Hexa-decafluorinated zinc phthalocyanine@f-MWCNT) on SPE	Cl_2	resistance/0.06 ppb	n.r.	n.r.	[63]
SbSI@CNTs/Au-microelectrode	ultrasonic bonding of SbSI@CNTs composite on Au-microelectrode	CO_2	amperometry/n.r.	n.r.	n.r.	[64]
MWCNT@polyaniline/FTO	drop coating of MWCNT@polyaniline nanocomposite on FTO	methanol vapor	resistance/≈50 ppm	n.r.	the signal remained almost constant for up to 20 days	[65]
Industrial by-products						
CuO-NP/MWCNT/GCE	electrodeposition of CuO-NP on MWCNT/GCE	glycerol	amperometry/5.8 × 10^{-6} g dm^{-3}	biodiesel samples	n.r.	[66]
Au-NP/Grf-Red@MWCNT/GCE	electrodeposition of Au-NP on Grf-Red@MWCNT/GCE	bisphenol A	DPV/1.0 × 10^{-9} mol L^{-1}	river water and shopping receipt samples	98% after 30 days	[67]
Pd-NP/Grf-Red@MWCNT/GCE	electrodeposition of Pd-NP on Grf-Red@MWCNT/GCE	hydrazine	amperometry/0.15 μmol L^{-1}	tap water spiked with hydrazine	n.r.	[68]
Prussian blue/CTS@MWCNT /GCE	electrodeposition of Prussian blue complex on GCE Modified with CTS@MWCNT nanocomposite	hydrogen peroxide	amperometry/0.10 μmol L^{-1}	routine analysis in pure electrolyte	90.5–92.6% after two weeks	[69]

Au-NP: gold nanoparticles; Ag-NP: silver nanoparticles; Cu-MP: copper microparticles; CuO-NP: copper oxide nanoparticles; Pd-NP: palladium nanoparticles; CTS: chitosan; Co-Pht: cobalt phthalocyanine; IL: ionic liquid; HPU: hydrothane polyurethane; β-CD: β-cyclodextrin; BiF: bismuth film; MIP: molecular imprinted polymer; MWCNT: multi-walled carbon nanotubes; f-MWCNT: functionalized multi-walled carbon nanotubes; 3D-Grf: three-dimensional graphene; Grf-Ox: graphene oxide; Grf-Red: reduced graphene; HIV-p24: retrovirus of the AIDS; GCE: glassy carbon electrode; SPE: screen-printed electrode; CPE: carbon paste electrode; FTO: fluorine doped thin oxide electrode; LSV: linear sweep voltammetry; CV: cyclic voltammetry; DPV: differential pulse voltammetry; SWV: square-wave voltammetry; ASV: adsorptive stripping voltammetry; SWAdSV: square-wave adsorptive stripping voltammetry; n.r.: no reported.

Appl. Sci. **2018**, *8*, 1925

5. Final Remarks

MWCNT and their (nano)composites have been the origin for exciting and versatile sensing platforms due to their inherent advantages, namely large active surface area, electrocatalytic properties, high and fast electron-transfer, signal enlargement, reduction of the overpotential, chemical inertness, among others. Their incorporation into electrochemical sensors has been improving the electroanalytical performance of these devices and simultaneously widening their scope and range of applications. However, some limitations related to the high cost of production of high purity MWCNT have been restricting their large-scale and industrial utilization. Thus, improvements on the current synthetic routes and purification methods or/and development of novel low cost and effective techniques are clearly needed. In addition, comprehensive characterization of the toxicity of MWCNT is another requirement to expand their application to in vivo assays. These achievements will contribute to the construction of reliable and reproducible MWCNT-based electrochemical sensors enlarging sensors exploitation in real in vivo monitoring and in on-line manufacturing applications.

Author Contributions: Conceptualization, T.M.B.F.O. and S.M.; Data curation, T.M.B.F.O.; Supervision, Simone Morais; Writing—original draft, T.M.B.F.O.; Writing—review and editing, S.M.

Funding: Simone Morais is grateful for financial support from the European Union (FEDER funds through COMPETE) and National Funds (Fundação para a Ciência e Tecnologia-FCT) through projects UID/QUI/50006/2013 and AAC No. 02/SAICT/2017—project No. 029547 "CECs(Bio)Sensing—(Bio)sensors for assessment of contaminants of emerging concern in fishery commodities", by FCT/MEC with national funds and co-funded by FEDER. She also thanks financial support by Norte Portugal Regional Operational Programme (NORTE 2020), under the PORTUGAL 2020 Partnership Agreement, through the European Regional Development Fund (ERDF): projects Norte-01-0145-FEDER-000011 and Norte-01-0145-FEDER-000024.

Acknowledgments: T.M.B.F. Oliveira thanks the Brazilian agencies CNPq, CAPES and FUNCAP for all scientific support to his projects with nanostructured materials.

Conflicts of Interest: The authors declare no conflict of interest.

References

1. Sudha, P.N.; Sangeetha, K.; Vijayalakshmi, K.; Barhoum, A. Nanomaterials history, classification, unique properties, production and market. In *Emerging Applications of Nanoparticles and Architecture Nanostructures— Current Prospects and Future Trends*; A Volume in Micro and Nano Technologies; Makhlouf, A.S.H., Barhoum, A., Eds.; Elsevier: Cambridge, MA, USA, 2018; pp. 341–384, ISBN 978-0-323-51254-1.

2. Soriano, M.S.; Zougagh, M.; Valcárcel, M.; Ríos, Á. Analytical Nanoscience and Nanotechnology: Where we are and where we are heading. *Talanta* **2018**, *177*, 104–121. [CrossRef] [PubMed]

3. Adams, F.C.; Barbante, C. Nanoscience, nanotechnology and spectrometry. *Spectrochim. Acta Part B* **2013**, *86*, 3–13. [CrossRef]

4. Liu, J.; Liu, L.; Lu, J.; Zhu, H. The formation mechanism of chiral carbon nanotubes. *Physica B* **2018**, *530*, 277–282. [CrossRef]

5. Kurkowska, M.; Awietjan, S.; Kozera, R.; Jezierska, E.; Boczkowska, A. Application of electroless deposition for surface modification of the multiwall carbon nanotubes. *Chem. Phys. Lett.* **2018**, *702*, 38–43. [CrossRef]

6. Zaporotskova, I.V.; Boroznina, N.P.; Parkhomenko, Y.N.; Kozhitov, L.V. Carbon nanotubes: Sensor properties. A review. *Mod. Electron. Mater.* **2016**, *2*, 95–105. [CrossRef]

7. Dumitrescu, I.; Unwin, P.R.; Macpherson, J.V. Electrochemistry at carbon nanotubes: Perspective and issues. *Chem. Commun.* **2009**, 6886–6901. [CrossRef] [PubMed]

8. Hamada, N.; Sawada, S.-I.; Oshiyama, A. New one-dimensional conductors: Graphitic microtubules. *Phys. Rev. Lett.* **1992**, *68*, 1579–1581. [CrossRef] [PubMed]

9. Mao, J.; Wang, Y.; Zhu, J.; Yu, J.; Hu, Z. Thiol functionalized carbon nanotubes: Synthesis by sulfur chemistry and their multi-purpose applications. *Appl. Surf. Sci.* **2018**, *447*, 235–243. [CrossRef]

10. Xiao, Z.; Elike, J.; Reynolds, A.; Moten, R.; Zhao, X. The fabrication of carbon nanotube electronic circuits with dielectrophoresis. *Microelectron. Eng.* **2016**, *164*, 123–127. [CrossRef]

11. Su, L.; Wang, X.; Wang, Y.; Zhang, Q. Roles of carbon nanotubes in novel energy storage devices. *Carbon* **2017**, *122*, 462–474. [CrossRef]

12. Guo, Y.; Shen, G.; Sun, X.; Wang, X. Electrochemical aptasensor based on multiwalled carbon nanotubes and graphene for tetracycline detection. *IEES Sens. J.* **2015**, *15*, 1951–1958. [CrossRef]

13. Liu, L.; Niu, Z.; Chen, J. Flexible supercapacitors based on carbon nanotubes. *Chin. Chem. Lett.* **2018**, *29*, 571–581. [CrossRef]

14. Parveen, S.; Kumar, A.; Husain, S.; Husain, M. Fowler Nordheim theory of carbon nanotube based field emitters. *Phys. B Condens. Matter.* **2017**, *505*, 1–8. [CrossRef]

15. Hulanicki, A.; Glab, S.; Ingman, F. Chemical sensors: Definitions and classification. *Pure Appl. Chem.* **1991**, *63*, 1247–1250. [CrossRef]

16. Kim, S.N.; Rusling, J.F.; Papadimitrakopoulos, F. Carbon nanotubes for electronic and electrochemical detection of biomolecules. *Adv. Mater.* **2007**, *19*, 3214–3228. [CrossRef] [PubMed]

17. López-Lorente, Á.; Valcárcel, M. The third way in analytical nanoscience and nanotechnology: Involvement of nanotools and nanoanalytes in the same analytical process. *Trends Analyt. Chem.* **2016**, *75*, 1–9. [CrossRef]

18. Harris, P.J. Engineering carbon materials with electricity. *Carbon* **2017**, *122*, 504–513. [CrossRef]

19. Abdalla, S.; Al-Marzouki, F.; Al-Ghamdi, A.A.; Abdel-Daiem, A. Different technical applications of carbon nanotubes. *Nanoscale Res. Lett.* **2015**, *10*, 1–10. [CrossRef] [PubMed]

20. Park, S.; Vosguerichian, M.; Zhenan Bao, Z. A review of fabrication and applications of carbon nanotube film-based flexible electronics. *Nanoscale* **2013**, *5*, 1727–1752. [CrossRef] [PubMed]

21. Rezaee, S.; Ghaderi, A.; Boochani, A.; Solaymani, S. Synthesis of multiwalled carbon nanotubes on Cu-Fe nano-catalyst substrate. *Res. Phys.* **2017**, *7*, 3640–3644. [CrossRef]

22. Dhore, V.G.; Rathod, W.S.; Patil, K.N. Synthesis and characterization of high yield multiwalled carbon nanotubes by ternary catalyst. *Mater. Today Proc.* **2018**, *5*, 3432–3437. [CrossRef]

23. Monthioux, M.; Serp, P.; Flahaut, E.; Razafinimanana, M.; Laurent, C.; Peigney, A.; Bacsa, W.; Broto, J.-M. Introduction to carbon nanotubes. In *Springer Handbook of Nanotechnology*, 2nd ed.; Bhushan, B., Ed.; Springer: Berlin, Germany, 2007; pp. 47–118, ISBN 3-540-29855-X.

24. Araga, R.; Sharma, C.S. One step direct synthesis of multiwalled carbon nanotubes from coconut shell derived charcoal. *Mater. Lett.* **2017**, *188*, 205–207. [CrossRef]

25. Rius, G.; Baldi, A.; Ziaie, B.; Atashbar, M.Z. Introduction to micro-/nanofabrication. In *Springer Handbook of Nanotechnology*, 4th ed.; Bhushan, B., Ed.; Springer: Berlin, Germany, 2017; pp. 51–86, ISBN 978-3-662-54355-9.

26. Yáñez-Sedeño, P.; Pingarrón, J.M.; Riu, J.; Rius, F.X. Electrochemical sensing based on carbon nanotubes. *Trends Anal. Chem.* **2010**, *29*, 939–953. [CrossRef]

27. Bandaru, P.R. Electrical properties and applications of carbon nanotube structures. *J. Nanosci. Nanotechnol.* **2007**, *7*, 1–29. [CrossRef]

28. Ahammad, A.J.S.; Lee, J.-J.; Rahman, M.A. Electrochemical sensors based on carbon nanotubes. *Sensors* **2009**, *9*, 2289–2319. [CrossRef] [PubMed]

29. Mao, A.; Li, H.; Yu, L.; Hu, X. Electrochemical sensor based on multi-walled carbon nanotubes and chitosan-nickel complex for sensitive determination of metronidazole. *J. Electroanal. Chem.* **2017**, *799*, 257–262. [CrossRef]

30. Holanda, L.F.; Ribeiro, F.W.P.; Sousa, C.P.; Casciano, P.N.S.; De Lima-Neto, P.; Correia, A.N. Multi-walled carbon nanotubes-cobalt phthalocyanine modified electrode for electroanalytical determination of acetaminophen. *J. Electroanal. Chem.* **2016**, *772*, 9–16. [CrossRef]

31. Montes, R.H.O.; Lima, A.P.; Cunha, R.R.; Guedes, T.J.; Dos Santos, W.T.P.; Nosso, E.; Richter, E.M.; Munoz, R.A.A. Size effects of multi-walled carbon nanotubes on the electrochemical oxidation of propionic acid derivative drugs: Ibuprofen and naproxen. *J. Electroanal. Chem.* **2016**, *775*, 9–16. [CrossRef]

32. Pavinatto, A.; Mercante, L.A.; Leandro, C.S.; Mattoso, L.H.C.; Correa, D.S. Layer-by-Layer assembled films of chitosan and multi-walled carbon nanotubes for the electrochemical detection of 17α-ethinylestradiol. *J. Electroanal. Chem.* **2015**, *755*, 215–220. [CrossRef]

33. Chen, L.; Li, K.; Zhu, H.; Meng, L.; Chen, J.; Li, M.; Zhu, Z. A chiral electrochemical sensor for propranolol based on multi-walled carbon nanotubes/ionic liquids nanocomposite. *Talanta* **2013**, *105*, 250–254. [CrossRef] [PubMed]

34. Hundari, F.F.; Souza, J.C.; Zanoni, M.V.B. Adsorptive stripping voltammetry for simultaneous determination of hydrochlorothiazide and triamterene in hemodialysis samples using a multi-walled carbon nanotube-modified glassy carbon electrode. *Talanta* **2018**, *179*, 652–657. [CrossRef] [PubMed]

35. Zhai, H.; Wang, H.; Wanh, S.; Chen, Z.; Wang, S.; Zhou, Q.; Pan, Y. Electrochemical determination of mangiferin and icariin based on Au-AgNPs/MWNTs-SGSs modified glassy carbon electrode. *Sens. Actuators B Chem.* **2018**, *255*, 1771–1780. [CrossRef]

36. Yang, X.; Yu, X.; Heng, Y.; Wang, F. Facile fabrication of 3D graphene-multi walled carbon nanotubes network and its use as a platform for natamycin detection. *J. Electroanal. Chem.* **2018**, *816*, 54–61. [CrossRef]

37. Deng, K.; Liu, X.; Li, C.; Hou, Z.; Huang, H. An electrochemical omeprazole sensor based on shortened multi-walled carbon nanotubes-Fe_3O_4 nanoparticles and poly(2,6-pyridinedicarboxylic acid). *Sens. Actuators B Chem.* **2017**, *253*, 1–9. [CrossRef]

38. Khaled, E.; Khalil, M.M.; El Aziz, G.M.A. Calixarene/carbon nanotubes based screen printed sensors for potentiometric determination of gentamicin sulphate in pharmaceutical preparations and spiked surface water samples. *Sens. Actuators B Chem.* **2017**, *244*, 876–884. [CrossRef]

39. Başkaya, G.; Yıldız, Y.; Savk, A.; Okyay, T.O.; Eriş, S.; Sert, H.; Şen, F. Rapid, sensitive, and reusable detection of glucose by highly monodisperse nickel nanoparticles decorated functionalized multi-walled carbon nanotubes. *Biosens. Bioelectron.* **2017**, *91*, 728–733. [CrossRef] [PubMed]

40. Wang, D.; Huang, B.; Liu, J.; Guo, X.; Abudukeyoumu, G.; Zhang, Y.; Ye, B.-C.; Li, Y. A novel electrochemical sensor based on Cu@Ni/MWCNTs nanocomposite for simultaneous determination of guanine and adenine. *Biosens. Bioelectron.* **2018**, *102*, 389–395. [CrossRef] [PubMed]

41. Li, J.; Lee, E.-C. Functionalized multi-wall carbon nanotubes as an efficient additive for electrochemical DNA sensor. *Sens. Actuators B Chem.* **2017**, *239*, 652–659. [CrossRef]

42. Ma, Y.; Shen, X.-L.; Zeng, Q.; Wang, H.-S.; Wang, L.-S. A multi-walled carbon nanotubes based molecularly imprinted polymers electrochemical sensor for the sensitive determination of HIV-p24. *Talanta* **2017**, *164*, 121–127. [CrossRef] [PubMed]

43. Anirudhan, T.S.; Alexander, S. A potentiometric sensor for the trace level determination of hemoglobin in real samples using multiwalled carbon nanotube based molecular imprinted polymer. *Eur. Polym. J.* **2017**, *97*, 84–93. [CrossRef]

44. Gutierrez, F.A.; Rubianes, M.D.; Rivas, G.A. Electrochemical sensor for amino acids and glucose based on glassy carbon electrodes modified with multi-walled carbon nanotubes and copper microparticles dispersed in polyethylenimine. *J. Electroanal. Chem.* **2016**, *765*, 16–21. [CrossRef]

45. Ji, J.; Zhou, Z.; Zhao, X.; Sun, J.; Sun, X. Electrochemical sensor based on molecularly imprinted film at Au nanoparticles-carbon nanotubes modified electrode for determination of cholesterol. *Biosens. Bioelectron.* **2015**, *66*, 590–595. [CrossRef] [PubMed]

46. Taurino, I.; Van Hoof, V.; De Micheli, G.; Carrara, S. Superior sensing performance of multi-walled carbon nanotube-based electrodes to detect unconjugated bilirubin. *Thin Solid Films* **2013**, *548*, 546–550. [CrossRef]

47. Sharma, V.V.; Gualandi, I.; Vlamidis, Y.; Tonelli, D. Electrochemical behavior of reduced graphene oxide and multi-walled carbon nanotubes composites for catechol and dopamine oxidation. *Electrochim. Acta* **2017**, *246*, 415–423. [CrossRef]

48. Li, J.; Sun, Q.; Mao, Y.; Bai, Z.; Ning, X.; Zheng, J. Sensitive and low-potential detection of NADH based on boronic acid functionalized multi-walled carbon nanotubes coupling with an electrocatalysis. *J. Electroanal. Chem.* **2017**, *794*, 1–7. [CrossRef]

49. Wayu, M.B.; DiPasquale, L.T.; Schwarzmann, M.A.; Gillespie, S.D.; Leopold, M.C. Electropolymerization of β-cyclodextrin onto multi-walled carbon nanotube composite films for enhanced selective detection of uric acid. *J. Electroanal. Chem.* **2016**, *783*, 192–200. [CrossRef]

50. Tarditto, L.V.; Arévalo, F.J.; Zon, M.A.; Ovando, H.G.; Vettorazzi, N.R.; Fernández, H. Electrochemical sensor for the determination of enterotoxigenic *Escherichia coli* in swine feces using glassy carbon electrodes modified with multi-walled carbon nanotubes. *Microchem. J.* **2016**, *127*, 220–225. [CrossRef]

51. Sipa, K.; Brycht, M.; Leniart, A.; Urbaniak, P.; Nosal-Wiercińska, A.; Pałecz, B.; Skrzypek, S. β-Cyclodextrins incorporated multi-walled carbon nanotubes modified electrode for the voltammetric determination of the pesticide dichlorophen. *Talanta* **2018**, *176*, 625–634. [CrossRef] [PubMed]

52. Özcan, A.; Gürbüz, M. Development of a modified electrode by using a nanocomposite containing acid-activated multi-walled carbon nanotube and fumed silica for the voltammetric determination of clopyralid. *Sens. Actuators B Chem.* **2018**, *255*, 262–267. [CrossRef]

53. Ghodsi, J.; Rafati, A.A. A voltammetric sensor for diazinon pesticide based on electrode modified with TiO_2 nanoparticles covered multi walled carbon nanotube nanocomposite. *J. Electroanal. Chem.* **2017**, *807*, 1–9. [CrossRef]

54. Wei, X.-P.; Luo, Y.-L.; Xu, F.; Chen, Y.-S.; Yang, L.H. In-situ non-covalent dressing of multi-walled carbon nanotubes@titanium dioxides with carboxymethyl chitosan nanocomposite electrochemical sensors for detection of pesticide residues. *Mater. Des.* **2016**, *111*, 445–452. [CrossRef]

55. Ertan, B.; Eren, T.; Ermiş, İ.; Saral, H.; Atar, N.; Yola, M.L. Sensitive analysis of simazine based on platinum nanoparticles on polyoxometalate/multi-walled carbon nanotubes. *J. Colloid. Interface Sci.* **2016**, *470*, 14–21. [CrossRef] [PubMed]

56. Xuan, X.; Park, J.Y. A miniaturized and flexible cadmium and lead ion detection sensorbased on micro-patterned reduced graphene oxide/carbonnanotube/bismuth composite electrodes. *Sens. Actuators B Chem.* **2018**, *255*, 1220–1227. [CrossRef]

57. Roushani, M.; Saedi, Z.; Hamdi, F.; Dizajdizi, B.Z. Preparation an electrochemical sensor for detection of manganese (II) ions using glassy carbon electrode modified with multi walled carbon nanotube-chitosan-ionic liquid nanocomposite decorated with ion imprinted polymer. *J. Electroanl. Chem.* **2017**, *804*, 1–6. [CrossRef]

58. Firmino, M.L.M.; Morais, S.; Correia, A.N.; De Lima-Neto, P.; Carvalho, F.A.O.; Castro, S.S.L.; Oliveira, T.M.B.F. Sensor based on β-NiO$_x$ hybrid film/multi-walled carbon nanotubes composite electrode for groundwater salinization inspection. *Chem. Eng. J.* **2017**, *323*, 47–55. [CrossRef]

59. Sudha, V.; Kumar, S.M.S.; Thangamuthu, R. Simultaneous electrochemical sensing of sulphite and nitrite on acid-functionalized multi-walled carbon nanotubes modified electrodes. *J. Alloys Compd.* **2018**, *749*, 990–999. [CrossRef]

60. Li, Q.; Zhang, Q.; Ding, L.; Zhou, D.; Cui, H.; Wei, Z.; Zhai, J. Synthesis of silver/multi-walled carbon nanotubes composite and its application for electrocatalytic reduction of bromate. *Chem. Eng. J.* **2013**, *217*, 28–33. [CrossRef]

61. Qiu, X.; Lu, L.; Leng, J.; Yu, Y.; Wang, W.; Jiang, M.; Bai, L. An enhanced electrochemical platform based on graphene oxide and multi-walled carbon nanotubes nanocomposite for sensitive determination of sunset yellow and tartrazine. *Food Chem.* **2019**, *190*, 889–895. [CrossRef] [PubMed]

62. Tang, J.; Jin, B. Poly (crystal violet)-Multi-walled carbon nanotubes modified electrode for electroanalytical determination of luteolin. *J. Electroanal. Chem.* **2016**, *780*, 46–52. [CrossRef]

63. Sharma, A.K.; Mahajan, A.; Bedi, R.K.; Kumar, S.; Debnath, A.K.; Aswal, D.K. Non-covalently anchored multi-walled carbon nanotubes with hexa-decafluorinated zinc phthalocyanine as ppb level chemiresistive chlorine sensor. *Appl. Surf. Sci.* **2018**, *427*, 202–209. [CrossRef]

64. Jesionek, M.; Nowak, M.; Mistewicz, K.; Kępińska, M.; Stróż, D.; Bednarczyk, I.; Paszkiewicz, R. Sonochemical growth of nanomaterials in carbon nanotube. *Ultrasonics* **2018**, *83*, 179–187. [CrossRef] [PubMed]

65. Bora, A.; Mohan, K.; Pegu, D.; Gohain, C.B.; Dolui, S.K. A room temperature methanol vapor sensor based on highlyconducting carboxylated multi-walled carbon nanotube/polyanilinenanotube composite. *Sens. Actuators B Chem.* **2017**, *253*, 977–986. [CrossRef]

66. Arévalo, F.J.; Osuna-Sánchez, Y.; Sandoval-Cortés, J.; Tocco, A.D.; Granero, A.M.; Robledo, S.N.; Zon, M.A.; Vettorazzi, N.R.; Martínez, J.L.; Segura, E.P.; et al. Development of an electrochemical sensor for the determination of glycerol based on glassy carbon electrodes modified with a copper oxide nanoparticles/multiwalled carbon nanotubes/pectin composite. *Sens. Actuators B Chem.* **2017**, *244*, 949–957. [CrossRef]

67. Yu, H.; Feng, X.; Chen, X.-X.; Qiao, J.-L.; Gao, X.-L.; Xu, B.; Gao, L.-J. Electrochemical determination of bisphenol A on a glassy carbon electrode modified with gold nanoparticles loaded on reduced graphene oxide-multi walled carbon nanotubes composite. *Chin. J. Anal. Chem.* **2017**, *45*, 713–720. [CrossRef]

68. Hu, J.; Zhao, Z.; Zhang, J.; Li, G.; Li, P.; Zhang, W.; Lian, K. Synthesis of palladium nanoparticle modified reduced graphene oxide and multi-walled carbon nanotube hybrid structures for electrochemical applications. *Appl. Surf. Sci.* **2017**, *396*, 523–529. [CrossRef]

69. Wang, C.; Zhang, K.; Zhang, N.; Zhang, L.; Wang, H.; Xu, J.; Shi, H.; Zhuo, X.; Qin, M.; Wu, X. A simple strategy for fabricating a prussian blue/chitosan/carbon nanotube composite and its application for the sensitive determination of hydrogen peroxide. *Micro Nano Lett.* **2016**, *12*, 23–26. [CrossRef]

applied sciences

MDPI

Article

Controlling the Dissolution Rate of Hydrophobic Drugs by Incorporating Carbon Nanotubes with Different Levels of Carboxylation

Kun Chen and Somenath Mitra *

Department of Chemistry and Environmental Science, New Jersey Institute of Technology, Newark, NJ 07102, USA; kc226@njit.edu
* Correspondence: somenath.Mitra@njit.edu; Tel.: +1-973-596-5611; Fax: +1-973-596-3586

Received: 12 March 2019; Accepted: 3 April 2019; Published: 9 April 2019

Abstract: We present the anti-solvent precipitation of hydrophobic drugs griseofulvin (GF) and sulfamethoxazole (SMZ) in the presence of carboxylated carbon nanotubes (f-CNTs). The aqueous dispersed f-CNTs were directly incorporated into the drug particles during the precipitation process. f-CNTs with different levels of carboxylation were tested where the hydrophilicity was varied by altering the C:COOH ratio. The results show that the hydrophilic f-CNTs dramatically enhanced the dissolution rate for both drugs, and the enhancement corresponded to the hydrophilicity of f-CNTs. The time to reach 80% dissolution (t80) reduced from 52.5 min for pure SMZ to 16.5 min when incorporated f-CNTs that had a C:COOH ratio of 23.2 were used, and to 11.5 min when the ratio dropped to 16. A corresponding decrease was observed for SMZ for the above-mentioned f-CNTs. The study clearly demonstrates that it is possible to control the dissolution rate of hydrophobic drugs by altering the level of carboxylation of the incorporated CNTs.

Keywords: hydrophobic drugs; drug delivery; functionalized carbon nanotubes; dissolution rate; nanomedicine

1. Introduction

Many drugs referred to as Class II and Class IV drugs have low solubility which limits their bioavailability and consequently their effectiveness as therapeutic agents [1]. The solubility and bioavailability are typically improved by particle size reduction, which is described by the Noyes Whitney equation [2]. Typically, micro and nano drug particles are formed via mechanical size reduction such as dry/wet milling and homogenization [3], and also via precipitation techniques [4]. Anti-solvent precipitation has been used to synthesize micro and nano particles of hydrophobic drugs [5,6]. Here, an antisolvent is used to precipitate crystals from a solution whose properties can be controlled by altering process conditions and the use of additives [7,8]. Dissolution rates of hydrophobic drugs have been enhanced by the addition of hydrophilic moieties to the formulation. For example, different cellulosic materials [9] have been used as co-precipitating agents and hydrophilic silica nanoparticles have been used to promote faster aqueous dissolution [10]. Various polymers have been employed as peptide carriers in diabetes, oncology, and cardiovascular drugs [11]. Solid dispersion is an increasingly popular method that uses hydroxypropyl methylcellulose (HPMC), polyvinylpyrrolidone (PVP), polyethylene glycol (PEG), and polymer micelles as carriers for insoluble drugs [12,13]. Glucosamine hydrochloride has been used in solid dispersions [14] and hydrophobic molecules have been included in cyclodextrin [15] to enhance dissolution rates.

A drug carrier can be directly incorporated into the drug crystal during anti-solvent precipitation, and the latter can play multiple roles. For example, it can serve as a nucleation site for crystal formation, provide colloidal stability during crystal formation, and be used as a drug delivery vehicle such as a targeting agent or one that alters bioavailability by changing the dissolution rate. It is well known that functionalization is an effective means with which to control aqueous behavior of nanotubes including colloidal stability as well as their solubilization capacity towards hydrophobic molecules [16–18]. Fiber-like carbon nanotubes (CNTs) can actually be incorporated into drug crystals, and if the functionalized CNTs) are hydrophilic, they can attract water molecules and bring them to drug crystals, leading up to faster dissolution. The hydrophilicity of functionalized CNTs can also be altered to change the dissolution rate, which is a phenomenon that can be used to control the release of the drug. One of the concerns for using CNTs in drug delivery and biomedical applications is the toxicity of oxidized CNTs. A number of in vitro and in vivo studies have been performed that show conflicting reports from both the type of CNTs and bioactivity of interest [19–27]. However, studies have shown that carboxylation is an effective way to reduce toxicity [28,29].

The unique properties of carbon nanotubes have led to various applications in biological and environmental fields [30,31]. There has been much interest regarding carbon nanotubes in nanomedicine and tissue engineering applications [32–39]. The CNTs have been used to deliver a wide range of small and large molecules for controlled release. Small drug molecules as well as peptides, vaccines, antibodies, nucleic acids, proteins, and genes have been attached to CNTs [40–43]. Targeted drug delivery using CNTs has been successful [44], and CNTs have shown permeability into tumor tissues via endocytosis [45].

The key to the applications of CNTs in drug delivery is their attachment to drug molecules. Different molecules/species can be attached to CNTs via covalent or non-covalent bonding. Covalent attachment to CNTs can provide secure loading of a molecule, and drugs such as paclitaxel, toxoid, doxorubicin, boron-bearing agents, methotrexate, and 10-hydroxycamptothecin have been linked to CNTs via non-biodegradable or degradable linkages [46,47]. If the drug is attached to the CNTs through a degradable linkage, the released drug's activity and functionality need to be maintained, which can be a challenge. On the other hand, non-covalent approaches do not cause changes in the chemistry of drugs. The non-covalent approach to drug loading is to load the molecule onto the CNT surface by simple adsorption, π-stacking, hydrophobic interaction, or capillarity-induced filling [48–50]. Both pure CNTs and functionalized CNTs have been used in drug delivery, and in the case of noncovalent bonding, the advantages of functionalized CNTs can still be utilized.

Among functionalized CNTs, carboxylated CNTs (f-CNTs) are highly water dispersible and our studies have demonstrated their potential to enhance dissolution rates [51]. F-CNTs can be synthesized so that the carbon to oxygen atomic ratio can be varied to give different levels of hydrophilicity and it is conceivable that by varying the degree of functionalization, the drug can be released at different rates. Therefore, the level of functionalization is expected to be an important factor. The objective of this work was to study the effect of the degree of functionalization of the incorporated f-CNTs on hydrophobic drugs during anti-solvent synthesis of micron-scale drug particles as well as the dissolution rates. Of particular interest to this study were the antifungal agent griseofulvin (GF) and the antibiotic sulfamethoxazole (SMZ).

2. Material and Methods

2.1. Materials

Sodium dodecyl sulphate (SDS) was purchased from GFS (G. Frederick Smith) Chemicals Inc (GFS Chemicals Inc, Powell, OH, USA), hydrochloride acid was purchased from Fisher Scientific (Thermo Fisher Scientific Inc., Waltham, MA, USA), and raw multiwall carbon nanotubes nanotube (20–30 nm diameter, 10–30 μm length, purity > 95 wt%) was purchased from Cheap Tubes (Cheap Tubes Inc., Grafton, VT, USA), while GF, SMZ, sulfuric acid (95–98%), and nitric acid (70%) were purchased from Sigma Aldrich (MilliporeSigma Corporate, St. Louis, MO, USA). Purified Milli-Q Plus water (MilliporeSigma Corporate, St. Louis, MO, USA) was used in all experiments.

2.2. Methods

Carboxylated multiwall carbon nanotubes were synthesized following a methodology published previously in [51]. CNTs were reacted with a mixture of concentrated sulfuric acid and nitric acid at 140 °C for 5, 10, and 40 minutes, respectively, in a microwave reactor (model: CEM Mars). This led to the formation of various amounts of carboxylic groups on the CNTs surface that had different hydrophilicity [28]. The carboxylated CNTs were filtered through a 10μm polytetrafluoroethylene (PTFE) membrane filter, washed to a neutral pH, and dried under vacuum at 65 °C.

Drug/CNT composites were prepared by anti-solvent precipitation at room temperature. GF or SMZ (Figure 1) was saturatively dissolved in acetone. An antisolvent was prepared by dispersing f-CNTs in water under sonication for 10 min. The antisolvent was added dropwise into the drug solution under sonication and the solution turned cloudy immediately after the addition of the f-CNT suspension, which indicated crystal formation of the f-CNT/drug composites (which are referred to as GF-CNT$_X$ and SMZ-CNT$_X$, respectively, where x represents the carbon to oxygen ratio). The resulting solution was filtered through a 10 μm PTFE membrane filter, washed, and dried in a vacuum oven until a constant weight was maintained.

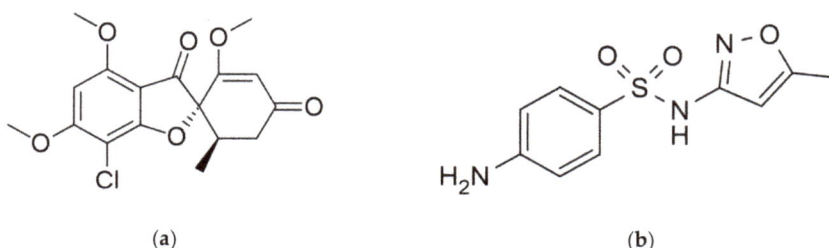

(a)   **(b)**

Figure 1. Structural formula for (**a**) griseofulvin (GF) and (**b**) sulfamethoxazole (SMZ).

The resulting drug/CNT composites were characterized with a scanning electron microscope (SEM), energy-dispersive X-ray spectroscopy (EDX), transmission electron microscopy (TEM), differential scanning calorimetry (DSC), an X-Ray diffractometer (XRD), thermogravimetric analysis (TGA), and elemental analysis. The dissolution was tested by dissolution testing apparatus 2. SEM and EDX was performed with a LEO 1530VP (LEO Electron Microscopy Inc., Thornwood, NY, USA) and JEOL JSM-7900F (JEOL Ltd., Tokyo, Japan). TEM was performed with a Hitachi H-7500 Tungsten/LaB6 (Hitachi, Ltd., Tokyo, Japan) with a 100 KV energy bean. TGA was performed with a Perkin Elmer Pyris 1 thermogravimetric analyzer (PerkinElmer Inc., Waltham, MA, USA) which heated the samples from 30 °C to 1200 °C at 10 °C/min in air. Elemental analysis was performed using a Perkin-Elmer 2400 Series II elemental analyzer (PerkinElmer Inc., Waltham, MA, USA). Raman spectroscopy was performed with a Thermo Scientific DXR Raman Microscope (Thermo Fisher Scientific Inc., Waltham, MA, USA) with a 532 nm filter. Melting points were measured with a Perkin Elmer DSC 6000 (PerkinElmer Inc., Waltham, MA, USA). DSC was carried out under nitrogen; GF-CNT$_X$ samples were heated from 30 °C to 250 °C at 20 °C /min while SMZ-CNT$_X$ samples were heated from 30 °C to 200 °C at 20 °C /min. XRD was performed with a PANalytical EMPYREAN XRD (Malvern Panalytical Inc., Westborough, MA, USA) with a Cu tube as the X-ray source. A Symphony 7100 dissolution system (Distek, Inc., North Brunswick, NJ, USA) was used to study the dissolution behavior of the drug composites using a standard United States Pharmacopeia (USP) method (USP 41). The relative standard deviation of three repeat dissolution tests were below 4%.

3. Results and Discussion

EDX was used to examine the carbon and oxygen percentages in the functionalized CNTs. The degree of functionalization is shown in Table 1. As functionalization time increased, oxygen percentage increased and the C:COOH ratio decreased. After 40 min of functionalization, the oxygen percentage did not increase significantly and therefore treatment times beyond 40 minutes were not studied. The oxygen content of the different f-CNTs varied from 6.1 to 13.6% while the C:COOH ratio could be as low as 16:1. The f-CNTs were labeled based on the C:COOH ratio.

Table 1. Analysis of carboxylated carbon nanotubes (f-CNTs) with different levels of functionalization.

Treatment Time (min)	% by Weight		C:COOH
	C	O	
0	92.7	6.1	39.5
5	89.7	9.9	23.2
10	87.9	11.2	19.8
40	86.3	13.6	16.0

The concentrations of f-CNTs in GF were calculated from the TGA (Figure 2). The amount of f-CNTs in the GF crystals prepared from f-CNT$_{23.2}$, f-CNT$_{19.8}$, and f-CNT$_{16.0}$ suspensions were found to be 3.9, 4.2, and 3.8%, respectively. The values were calculated based on the weight percent at the temperatures from which f-CNTs started to burn out (around 300 °C) for each composite minus the corresponding weight percent of pure GF at the same temperature. Since the behavior of the f-CNTs was similar to SMZ during thermogravimetric analysis, it was difficult to predict the composition of the SMZ-CNT samples by TGA. The concentrations of f-CNTs in the SMZ-CNTs were measured based on elemental analysis. The sulfur content was used to calculate the amount of SMZ in the composite, from which the amount of f-CNTs could be predicted. The amount of f-CNTs in the SMZ crystals prepared from f-CNT$_{23.2}$, f-CNT$_{19.8}$, and f-CNT$_{16.0}$ suspensions were found to be 2.9, 1.3, and 1.4%, respectively. These samples are referred to as SMZ-CNT$_X$ or GF-CNT$_X$, where x is the C:COOH ratio. It appears that the degree of functionalization did not significantly affect the weight percent of f-CNTs in the drug crystals formed during the anti-solvent precipitation.

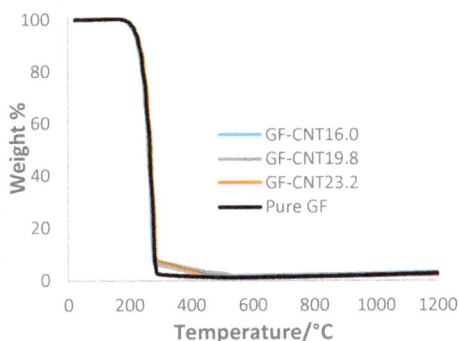

Figure 2. Thermogravimetric analysis (TGA) of GF-CNT$_X$.

The GF-CNT and SMZ-CNT sample morphologies were studied using SEM. Figure 3a–c show SEM images of GF-CNT$_{23.2}$, GF-CNT$_{19.8}$, and GF-CNT$_{16}$ at 25 K magnification. Figure 3d–f show SEM images of SMZ-CNT$_{23.2}$, SMZ-CNT$_{19.8}$ and SMZ-CNT$_{16}$ at the same magnification. The SEM images show that the f-CNTs were present on the crystal surface. Compared to the SEM images of pure GF and pure SMZ in Figure 3g,h, the crystal shape and size did not depend on CNT functionalization or incorporation. The TEM images (Figure 3i,j) show that the f-CNTs were also partially embedded in the drug crystals.

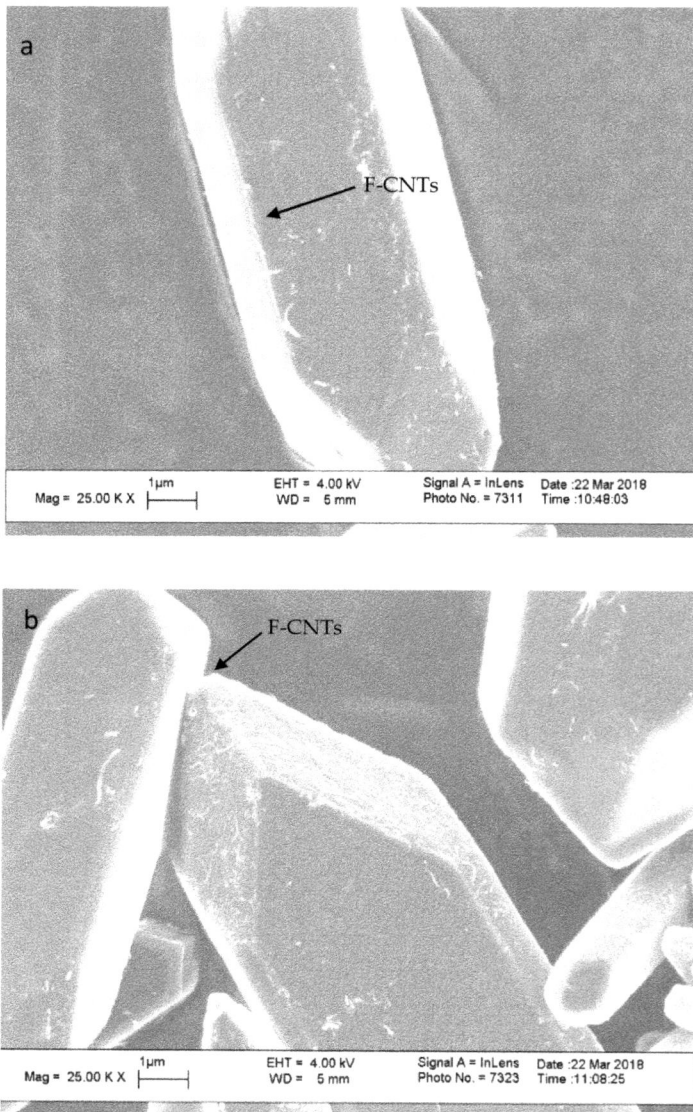

Figure 3. *Cont.*

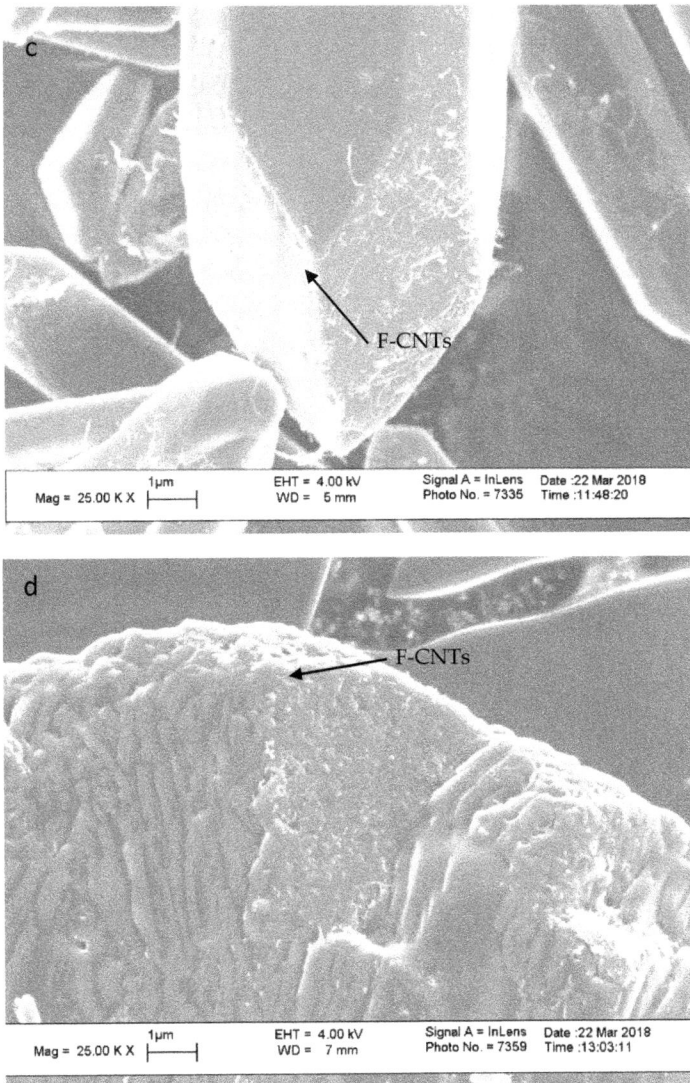

Figure 3. *Cont.*

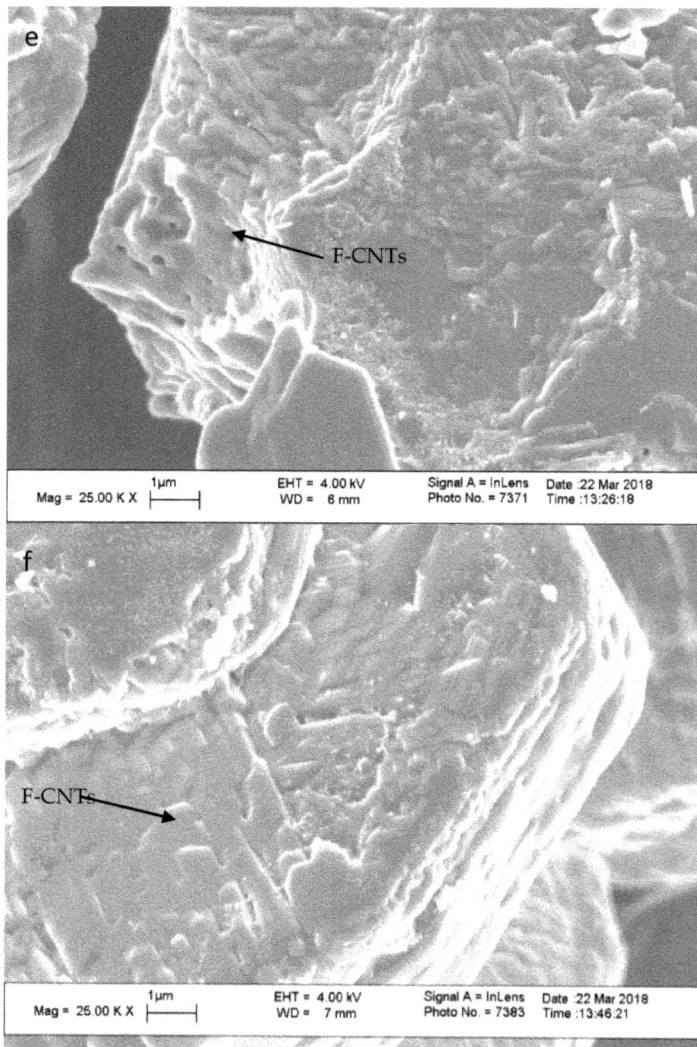

Figure 3. *Cont.*

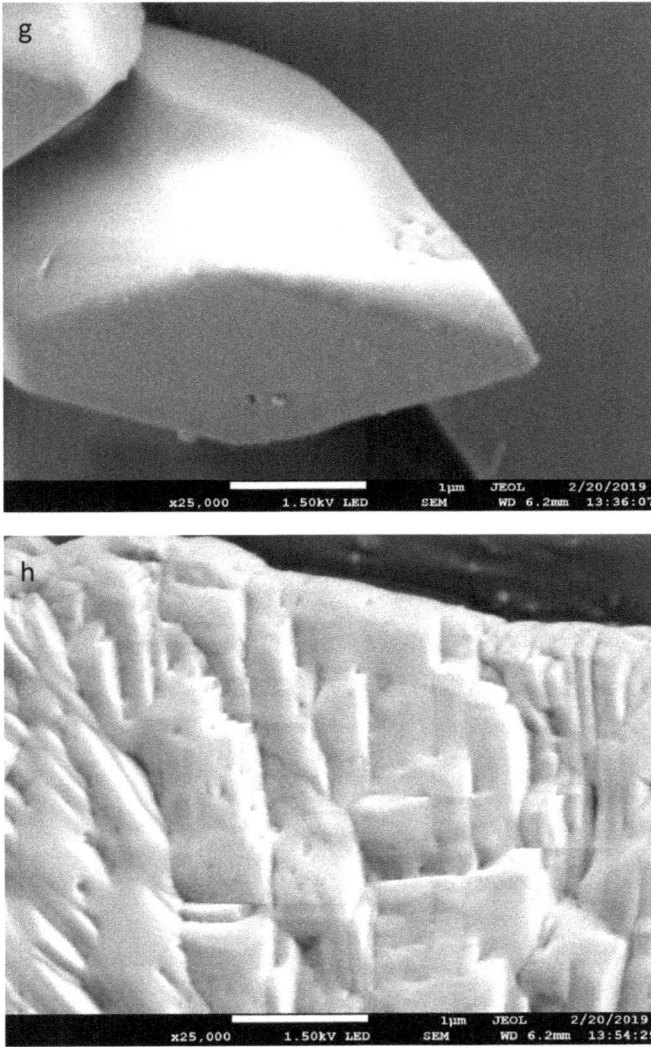

Figure 3. *Cont.*

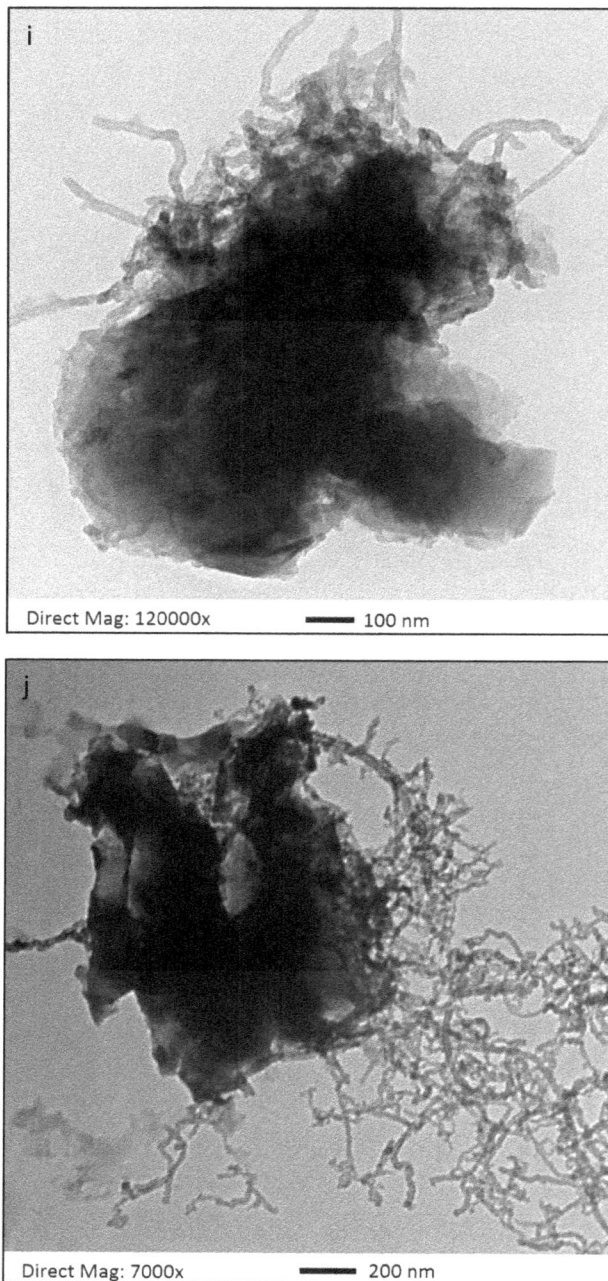

Figure 3. Scanning electron microscope (SEM) images of (**a**) GF-CNT$_{23.2}$, (**b**) GF-CNT$_{19.8}$, (**c**) GF-CNT$_{16}$, (**d**) SMZ-CNT$_{23.2}$, (**e**) SMZ-CNT$_{19.8}$, (**f**) SMZ-CNT$_{16}$, (**g**) pure GF, (**h**) pure SMZ, and Transmission electron microscopy (TEM) images of (**i**) GF-CNT$_{16}$, (**j**) SMZ-CNT$_{16}$.

Figure 4a shows the Raman spectra of f-CNTs, pure GF, and GF-CNTs with various degrees of functionalization. The typical spectral features of f-CNTs were overlaid with peaks from GF. The Raman spectra for pure GF and GF-CNT composites remained the same, indicating that the presence of the f-CNTs didn't change the chemical nature of the GF or its polymorphism, which are important considerations in drug development. A similar observation was found in Figure 4b, which shows Raman spectra of f-CNTs, pure SMZ, and SMZ-CNTs with various degree of functionalization.

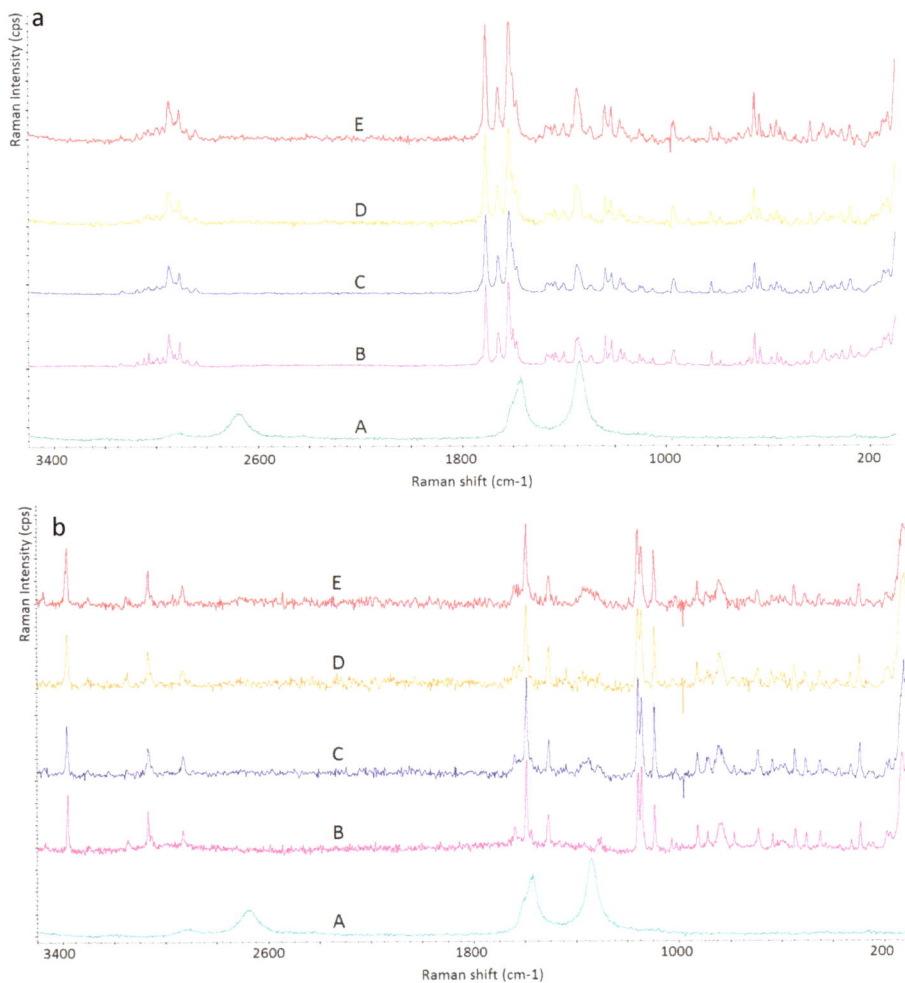

Figure 4. (**a**) Raman spectra of (A) f-CNT$_{23.2}$, (B) pure GF, (C) GF-CNT$_{23.2}$, (D) GF-CNT$_{19.8}$, and (E) GF-CNT$_{16}$; (**b**) Raman spectrum of (A) f-CNT$_{23.2}$, (B) pure SMZ, (C) SMZ-CNT$_{23.2}$, (D) SMZ-CNT$_{19.8}$, and (E) SMZ-CNT$_{16}$.

Crystal structures of GF-CNT$_x$ and SMZ-CNT$_x$ were also studied using XRD, and Figure 5 shows the diffractogram of GF-CNT$_x$ and SMZ-CNT$_x$. It can be seen that the crystal structure remained unchanged with the incorporation of the f-CNTs. The diffractograms of pure drugs and drug-CNTs were identical. This implies that there were no changes in polymorphism.

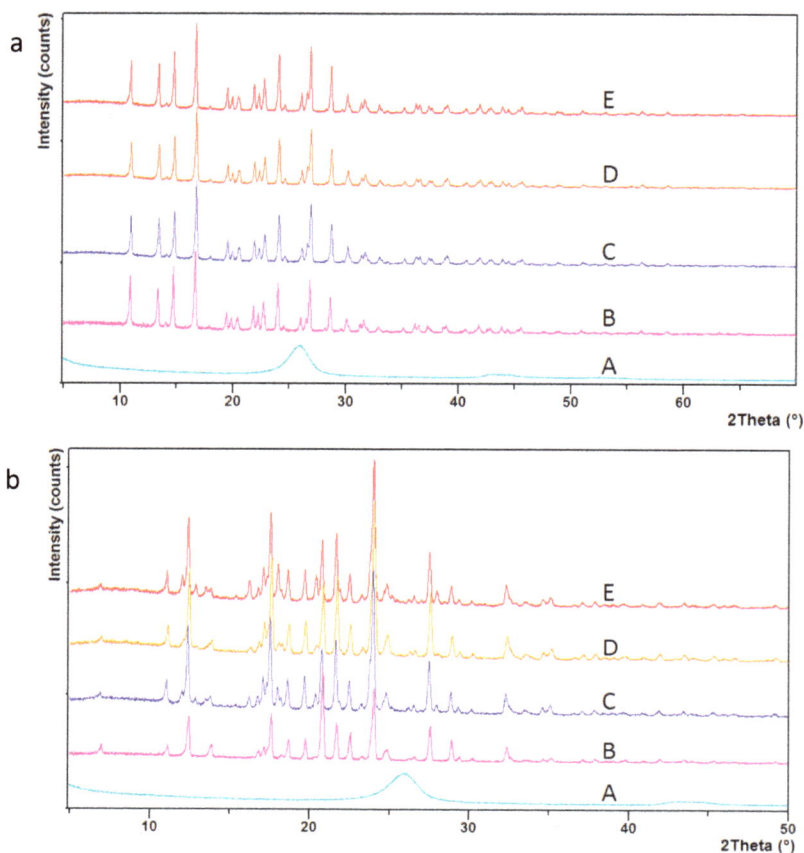

Figure 5. (**a**) X-Ray diffractometer (XRD) diffractogram of (A) f-CNT$_{23.2}$, (B) pure GF, (C) GF-CNT$_{23.2}$, (D) GF-CNT$_{19.8}$, and (E) GF-CNT$_{16}$; (**b**) XRD diffractogram of (A) f-CNT$_{23.2}$, (B) pure SMZ, (C) SMZ-CNT$_{23.2}$, (D) SMZ-CNT$_{19.8}$, and (E) SMZ-CNT$_{16}$.

The melting points of drug-CNT composites were analyzed by DSC 6000. The results are presented in Table 2. It can be seen that there was no significant change in melting points between the pure drug and its f-CNT composites.

Table 2. Dissolution and melting point of sulfamethoxazole-CNTs (SMZ-CNTs) and griseofulvin-CNTs (GF-CNTs).

	Incorporation %	C:COOH	T$_{50}$ (min)	T$_{80}$ (min)	M$_p$ (°C)
GF-CNT$_x$	0	39.5	8.0	>120.0	221.25
	3.9	23.2	6.0	60.0	220.75
	4.2	19.8	4.5	44.0	220.92
	3.8	16	4.0	30.5	221.01
SMZ-CNT$_x$	0	39.5	23.5	52.5	170.37
	2.9	23.2	8.5	16.5	170.21
	1.3	19.8	7.5	15.0	170.06
	1.4	16	6.0	11.5	170.05

Dissolution measurements were carried out based on the standard US Pharmacopeia Method (USP 41). GF-CNT composites were added to 4 mg/ml sodium dodecyl sulfate while SMZ-CNT composites were added to 0.1 N hydrochloric acid. The samples were stirred with a paddle at 75 rpm and heated to maintain a temperature of 37 ∘C. A small amount of medium was withdrawn at different times, filtered with a PTFE membrane to remove f-CNTs, and analyzed with Agilent 8453 UV-visible Spectroscopy System (Agilent, Santa Clara, CA, USA) to determine the amount of drug dissolved, at 291 nm for GF and 265 nm for SMZ. The dissolution data is presented in Figure 6.

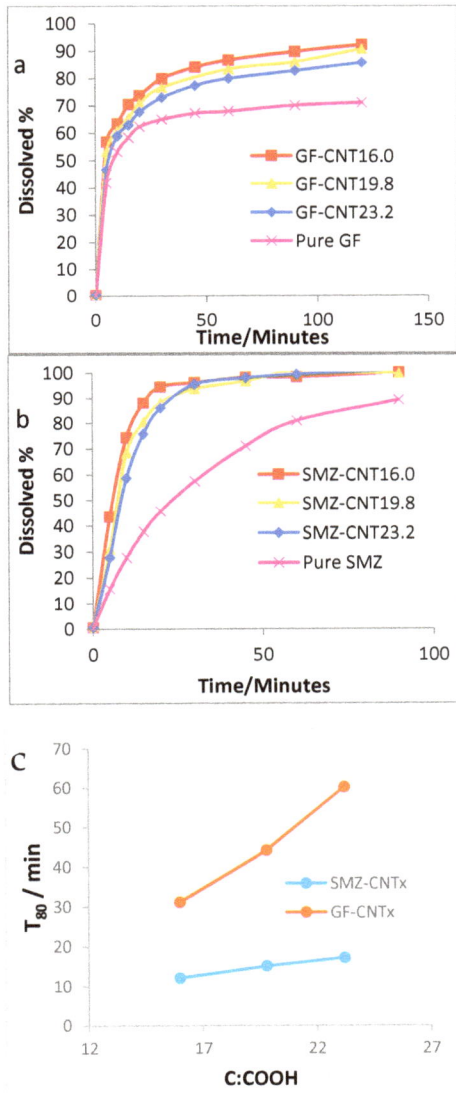

Figure 6. (**a**) Dissolution of GF-CNTs, (**b**) dissolution of SMZ-CNTs, and (**c**) time to reach 80% dissolution for drugs (t80) with a degree of functionalization of CNTs.

It is evident from both profiles that the increase in the level of functionalization in the f-CNTs enhanced the release of the drugs. The f-CNTs were hydrophilic and increased contact between the water and the drug crystals. The water molecules adsorbed on the hydrophilic carboxylic groups and then used these as a conduit to the drug crystal to increase dissolution.

The time necessary to reach 50% (t_{50}) and 80% (t_{80}) dissolution reduced with the incorporation of the f-CNTs. For GF samples, t_{50} and t_{80} reduced by as much as 50% and 75%, while for the SMZ samples, t_{50} and t_{80} were reduced by as much as 74% and 78%. As the C:COOH ratio decreased from 23.2 to 16, the t_{50} of GF dropped from 6.0 to 4.0 min, a reduction of 33%, and the t_{80} of GF dropped from 60.0 to 30.5 min, a reduction of 49%. The corresponding drop for SMZ was from 8.5 to 6.0 min, a reduction of 31% for t_{50} and from 16.5 to 11 min, a reduction of 33% for t_{80}.

4. Conclusions

The CNTs were oxidized to form f-CNTs with different levels of carboxylation. The SEM and TEM images show CNT incorporation into the drug crystals, and their presence was seen inside as well as outside the crystals. Based on Raman, XRD, and DSC measurements, it was evident that the f-CNTs did not change the crystal structure or the melting point. The dissolution rate was significantly enhanced with the incorporation of f-CNTs. By decreasing the C:COOH ratio in the f-CNTs, dissolution rates increased. Additionally, higher levels carboxylation showed lower values of these parameters. Therefore, it is evident that by varying the level of carboxylation, it is possible to control the dissolution rate of the hydrophobic drugs. This paper presents a novel approach to controlling the release rate by altering the level of CNT carboxylation. It is also possible that the degree of carboxylation can be used to alter other aspects of drug delivery, such as targeting capabilities.

Author Contributions: Conceptualization, K.C. and S.M.; methodology, K.C. and S.M.; software, K.C.; formal analysis, K.C.; investigation, K.C.; data curation, K.C.; writing—original draft preparation, K.C.; writing—review and editing, S.M.; project administration, S.M.; and funding acquisition, S.M.

Funding: This research was funded by a grant from the National Institute of Environmental Health Sciences (NIEHS) under Grant No. R01ES023209. Any opinions, findings, and conclusions or recommendations expressed in this material are those of the author(s) and do not necessarily reflect the views of the NIEHS.

Conflicts of Interest: The authors declare no conflict of interest.

References

1. Thorat, A.A.; Dalvi, S.V. Liquid antisolvent precipitation and stabilization of nanoparticles of poorly water soluble drugs in aqueous suspensions: Recent developments and future perspective. *Chem. Eng. J.* **2012**, *181–182*, 1–34. [CrossRef]
2. Xia, D.; Quan, P.; Piao, H.; Piao, H.; Sun, S.; Yin, Y.; Cui, F. Preparation of stable nitrendipine nanosuspensions using the precipitation–ultrasonication method for enhancement of dissolution and oral bioavailability. *Eur. J. Pharm. Sci.* **2010**, *40*, 325–334. [CrossRef] [PubMed]
3. Ryu, T.K.; Kim, S.E.; Kim, J.H.; Moon, S.K.; Choi, S.W. Biodegradable uniform microspheres based on solid-in-oil-in-water emulsion for drug delivery: A comparison of homogenization and fluidic device. *J. Bioact. Compat. Polym.* **2014**, *29*, 445–457. [CrossRef]
4. Kakran, M.; Sahoo, N.G.; Tan, I.L.; Li, L. Preparation of nanoparticles of poorly water-soluble antioxidant curcumin by antisolvent precipitation methods. *J. Nanoparticle Res.* **2012**, *14*, 757. [CrossRef]
5. Wang, Y.; Zhu, L.H.; Chen, A.Z.; Xu, Q.; Hong, Y.J.; Wang, S.B. One-Step Method to Prepare PLLA Porous Microspheres in a High-Voltage Electrostatic Anti-Solvent Process. *Materials* **2016**, *9*, 368. [CrossRef]
6. Imchalee, R.; Charoenchaitrakool, M. Gas anti-solvent processing of a new sulfamethoxazole−l-malic acid cocrystal. *J. Ind. Eng. Chem.* **2015**, *25*, 12–15. [CrossRef]
7. Singh, C.; Tiwari, V.; Mishra, C.P.; Shankar, R.; Sharma, D.; Jaiswal, S. Fabrication of cefpodoxime proxetil nanoparticles by solvent anti-solvent precipitation method for enhanced dissolution. *Int. J. Res. Pharm. Nano Sci.* **2015**, *4*, 18.
8. Park, M.W.; Yeo, S.D. Antisolvent crystallization of carbamazepine from organic solutions. *Chem. Eng. Res. Des.* **2012**, *90*, 2202–2208. [CrossRef]

9. Pachuau, L. *Application of Nanocellulose for Controlled Drug Delivery*; Wiley-VCH: Weinheim, Germany, 2017. [CrossRef]

10. Bharti, C.; Nagaich, U.; Pal, A.; Gulati, N. Mesoporous silica nanoparticles in target drug delivery system: A review. *Int. J. Pharm. Investig.* **2015**, *5*, 124–133. [CrossRef]

11. Du, A.W.; Stenzel, M.H. Drug Carriers for the Delivery of Therapeutic Peptides. *Biomacromolecules* **2014**, *15*, 1097–1114. [CrossRef]

12. Sim, T.; Lim, C.; Lee, J.W.; Kim, D.W.; Kim, Y.; Kim, M.; Choi, S.; Choi, H.G.; Lee, E.S.; Kim, K.S.; et al. Characterization and pharmacokinetic study of itraconazole solid dispersions prepared by solvent-controlled precipitation and spray-dry methods. *J. Pharm. Pharmacol.* **2017**, *69*, 1707–1715. [CrossRef]

13. Kalepu, S.; Nekkanti, V. Insoluble drug delivery strategies: Review of recent advances and business prospects. *Acta Pharm. Sin. B* **2015**, *5*, 442–453. [CrossRef]

14. Al-Hamidi, H.; Edwards, A.A.; Mohammad, M.A.; Nokhodchi, A. To enhance dissolution rate of poorly water-soluble drugs: Glucosamine hydrochloride as a potential carrier in solid dispersion formulations. *Colloids Surf. B Biointerfaces* **2010**, *76*, 170–178. [CrossRef]

15. Loftsson, T.; Brewster, M.E. Pharmaceutical Applications of Cyclodextrins. 1. Drug Solubilization and Stabilization. *J. Pharm. Sci.* **1996**, *85*, 1017–1025. [CrossRef]

16. Cavallaro, G.; Grillo, I.; Gradzielski, M.; Lazzara, G. Structure of Hybrid Materials Based on Halloysite Nanotubes Filled with Anionic Surfactants. *J. Phys. Chem. C* **2016**, *120*, 13492–13502. [CrossRef]

17. Lazzara, G.; Cavallaro, G.; Panchal, A.; Fakhrullin, R.; Stavitskaya, A.; Vinokurov, V.; Lvov, Y. An assembly of organic-inorganic composites using halloysite clay nanotubes. *Curr. Opin. Colloid Interface Sci.* **2018**, *35*, 42–50. [CrossRef]

18. Lisuzzo, L.; Cavallaro, G.; Lazzara, G.; Milioto, S.; Parisi, F.; Stetsyshyn, Y. Stability of Halloysite, Imogolite, and Boron Nitride Nanotubes in Solvent Media. *Appl. Sci.* **2018**, *8*, 1068. [CrossRef]

19. Liu, X.; Gurel, V.; Morris, D.; Murray, D.W.; Zhitkovich, A.; Kane, A.B.; Hurt, R.H. Bioavailability of nickel in single-wall carbon nanotubes. *Adv. Mater.* **2007**, *19*, 2790–2796. [CrossRef]

20. Poland, C.A.; Duffin, R.; Kinloch, I.; Maynard, A.; Wallace, W.A.H.; Seaton, A.; Stone, V.; Brown, S.; MacNee, W.; Donaldson, K. Carbon nanotubes introduced into the abdominal cavity of mice show asbestos-like pathogenicity in a pilot study. *Nat. Nanotechnol.* **2008**, *3*, 423. [CrossRef]

21. Sato, Y.; Yokoyama, A.; Shibata, K.I.; Akimoto, Y.; Ogino, S.I.; Nodasaka, Y.; Kohgo, T.; Tamura, K.; Akasaka, T.; Uo, M.; et al. Influence of length on cytotoxicity of multi-walled carbon nanotubes against human acute monocytic leukemia cell line THP-1 in vitro and subcutaneous tissue of rats in vivo. *Mol. Biosyst.* **2005**, *1*, 176–182. [CrossRef]

22. Jia, G.; Wang, H.; Yan, L.; Wang, X.; Pei, R.; Yan, T.; Zhao, Y.; Guo, X. Cytotoxicity of carbon nanomaterials: Single-wall nanotube, multi-wall nanotube, and fullerene. *Environ. Sci. Technol.* **2005**, *39*, 1378–1383. [CrossRef]

23. Pacurari, M.; Yin, X.J.; Zhao, J.; Ding, M.; Leonard, S.S.; Schwegler-Berry, D.; Ducatman, B.S.; Sbarra, D.; Hoover, M.D.; Castranova, V.; et al. Raw single-wall carbon nanotubes induce oxidative stress and activate MAPKs, AP-1, NF-kappaB, and Akt in normal and malignant human mesothelial cells. *Environ. Health Perspect.* **2008**, *116*, 1211–1217. [CrossRef]

24. Wang, L.; Luanpitpong, S.; Castranova, V.; Tse, W.; Lu, Y.; Pongrakhananon, V.; Rojanasakul, Y. Carbon nanotubes induce malignant transformation and tumorigenesis of human lung epithelial cells. *Nano Lett.* **2011**, *11*, 2796–2803. [CrossRef]

25. Pulskamp, K.; Diabaté, S.; Krug, H.F. Carbon nanotubes show no sign of acute toxicity but induce intracellular reactive oxygen species in dependence on contaminants. *Toxicol. Lett.* **2007**, *168*, 58–74. [CrossRef]

26. Kagan, V.E.; Konduru, N.V.; Feng, W.; Allen, B.L.; Conroy, J.; Volkov, Y.; Vlasova, I.I.; Belikova, N.A.; Yanamala, N.; Kapralov, A.; et al. Carbon nanotubes degraded by neutrophil myeloperoxidase induce less pulmonary inflammation. *Nat. Nanotechnol.* **2010**, *5*, 354. [CrossRef]

27. Takagi, D.; Homma, Y.; Hibino, H.; Suzuki, S.; Kobayashi, Y. Single-walled carbon nanotube growth from highly activated metal nanoparticles. *Nano Lett.* **2006**, *6*, 2642–2645. [CrossRef]

28. Wu, Z.; Wang, Z.; Yu, F.; Thakkar, M.; Mitra, S. Variation in chemical, colloidal and electrochemical properties of carbon nanotubes with the degree of carboxylation. *J. Nanoparticle Res.* **2017**, *19*, 16. [CrossRef]

29. Allegri, M.; Perivoliotis, D.K.; Bianchi, M.G.; Chiu, M.; Pagliaro, A.; Koklioti, M.A.; Trompeta, A.-F.A.; Bergamaschi, E.; Bussolati, O.; Charitidis, C.A. Toxicity determinants of multi-walled carbon nanotubes: The relationship between functionalization and agglomeration. *Toxicol. Rep.* **2016**, *3*, 230–243. [CrossRef]

30. Liu, P.; Cottrill, A.L.; Kozawa, D.; Koman, V.B.; Parviz, D.; Liu, A.T.; Yang, J.; Tran, T.Q.; Wong, M.H.; Wang, S.; et al. Emerging trends in 2D nanotechnology that are redefining our understanding of "Nanocomposites". *Nano Today* **2018**, *21*, 18–40. [CrossRef]

31. Khoshnevis, H.; Mint, S.M.; Yedinak, E.; Tran, T.Q.; Zadhoush, A.; Youssefi, M.; Pasquali, M.; Duong, H.M. Super high-rate fabrication of high-purity carbon nanotube aerogels from floating catalyst method for oil spill cleaning. *Chem. Phys. Lett.* **2018**, *693*, 146–151. [CrossRef]

32. Macchione, M.; Biglione, C.; Strumia, M. Design, Synthesis and Architectures of Hybrid Nanomaterials for Therapy and Diagnosis Applications. *Polymers* **2018**, *10*, 527. [CrossRef]

33. Fabbro, C.; Ali-Boucetta, H.; Ros, T.D.; Kostarelos, K.; Bianco, A.; Prato, M. Targeting carbon nanotubes against cancer. *Chem. Commun.* **2012**, *48*, 3911–3926. [CrossRef]

34. Bhise, K.; Sau, S.; Alsaab, H.; Kashaw, S.K.; Tekade, R.K.; Iyer, A.K. Nanomedicine for cancer diagnosis and therapy: Advancement, success and structure–activity relationship. *Ther. Deliv.* **2017**, *8*, 1003–1018. [CrossRef]

35. Martinelli, V.; Cellot, G.; Toma, F.M.; Long, C.S.; Caldwell, J.H.; Zentilin, L.; Giacca, M.; Turco, A.; Prato, M.; Ballerini, L.; et al. Carbon nanotubes promote growth and spontaneous electrical activity in cultured cardiac myocytes. *Nano Lett.* **2012**, *12*, 1831–1838. [CrossRef]

36. Vashist, A.; Kaushik, A.; Vashist, A.; Sagar, V.; Ghosal, A.; Gupta, Y.K.; Ahmad, S.; Nair, M. Advances in Carbon Nanotubes–Hydrogel Hybrids in Nanomedicine for Therapeutics. *Adv. Healthc. Mater.* **2018**, *7*, 1701213. [CrossRef]

37. Li, X.; Liu, H.; Niu, X.; Yu, B.; Fan, Y.; Feng, Q.; Cui, F.Z.; Watari, F. The use of carbon nanotubes to induce osteogenic differentiation of human adipose-derived MSCs in vitro and ectopic bone formation in vivo. *Biomaterials* **2012**, *33*, 4818–4827. [CrossRef]

38. Zhou, F.; Wu, S.; Yuan, Y.; Chen, W.R.; Xing, D. Mitochondria-targeting photoacoustic therapy using single-walled carbon nanotubes. *Small* **2012**, *8*, 1543–1550. [CrossRef]

39. Lee, H.J.; Park, J.; Yoon, O.J.; Kim, H.W.; Lee, D.Y.; Kim, D.H.; Lee, W.B.; Lee, N.E.; Bonventre, J.V.; Kim, S.S. Amine-modified single-walled carbon nanotubes protect neurons from injury in a rat stroke model. *Nat. Nanotechnol.* **2011**, *6*, 121–125. [CrossRef]

40. Pantarotto, D.; Partidos, C.D.; Hoebeke, J.; Brown, F.; Kramer, E.; Briand, J.P.; Muller, S.; Prato, M.; Bianco, A. Immunization with peptide-functionalized carbon nanotubes enhances virus-specific neutralizing antibody responses. *Chem. Biol.* **2003**, *10*, 961–966. [CrossRef]

41. Pantarotto, D.; Partidos, C.D.; Graff, R.; Hoebeke, J.; Briand, J.-P.; Prato, M.; Bianco, A. Synthesis, structural characterization, and immunological properties of carbon nanotubes functionalized with peptides. *J. Am. Chem. Soc.* **2003**, *125*, 6160–6164. [CrossRef]

42. McDevitt, M.R.; Chattopadhyay, D.; Kappel, B.J.; Jaggi, J.S.; Schiffman, S.R.; Antczak, C.; Njardarson, J.T.; Brentjens, R.; Scheinberg, D.A. Tumor targeting with antibody-functionalized, radiolabeled carbon nanotubes. *J. Nucl. Med.* **2007**, *48*, 1180–1189. [CrossRef]

43. Podesta, J.E.; Al-Jamal, K.T.; Herrero, M.A.; Tian, B.; Ali-Boucetta, H.; Hegde, V.; Bianco, A.; Prato, M.; Kostarelos, K. Antitumor activity and prolonged survival by carbon-nanotube-mediated therapeutic sirna silencing in a human lung xenograft model. *Small* **2009**, *5*, 1176–1185. [CrossRef]

44. Wulan, P.P.; Wulandari, H.; Ulwan, S.H.; Purwanto, W.W.; Mulia, K. Modification of carbon nanotube's dispersion using cetyltrimethyl ammonium bromide (CTAB) as cancer drug delivery. *Aip Conf. Proc.* **2018**, *1933*, 030008.

45. Lacerda, L.; Russier, J.; Pastorin, G.; Herrero, M.A.; Venturelli, E.; Dumortier, H.; Al-Jamal, K.T.; Prato, M.; Kostarelos, K.; Bianco, A. Translocation mechanisms of chemically functionalised carbon nanotubes across plasma membranes. *Biomaterials* **2012**, *33*, 3334–3343. [CrossRef]

46. Rezaei, S.J.T.; Hesami, A.; Khorramabadi, H.; Amani, V.; Malekzadeh, A.M.; Ramazani, A.; Niknejad, H. Pt(II) complexes immobilized on polymer-modified magnetic carbon nanotubes as a new platinum drug delivery system. *Appl. Organomet. Chem.* **2018**, *32*, e4401. [CrossRef]

47. Li, R.; Wu, R.; Zhao, L.; Hu, Z.; Guo, S.; Pan, X.; Zou, H. Folate and iron difunctionalized multiwall carbon nanotubes as dual-targeted drug nanocarrier to cancer cells. *Carbon* **2011**, *49*, 1797–1805. [CrossRef]

48. Tan, J.; Saifullah, B.; Kura, A.; Fakurazi, S.; Hussein, M. Incorporation of Levodopa into Biopolymer Coatings Based on Carboxylated Carbon Nanotubes for pH-Dependent Sustained Release Drug Delivery. *Nanomaterials* **2018**, *8*, 389. [CrossRef]
49. Chae, S.; Kim, D.; Lee, K.J.; Lee, D.; Kim, Y.O.; Jung, Y.C.; Rhee, S.D.; Kim, K.R.; Lee, J.O.; Ahn, S.; et al. Encapsulation and Enhanced Delivery of Topoisomerase I Inhibitors in Functionalized Carbon Nanotubes. *Acs Omega* **2018**, *3*, 5938–5945. [CrossRef]
50. Tan, J.M.; Arulselvan, P.; Fakurazi, S.; Ithnin, H.; Hussein, M.Z. A review on characterizations and biocompatibility of functionalized carbon nanotubes in drug delivery design. *J. Nanomater.* **2014**, *2014*, 20. [CrossRef]
51. Chen, K.; Mitra, S. Incorporation of functionalized carbon nanotubes into hydrophobic drug crystals for enhancing aqueous dissolution. *Colloids Surf. B Biointerfaces* **2019**, *173*, 386–391. [CrossRef]

applied
sciences

MDPI

Article

Effect of Carbon Nanotubes on Chloride Penetration in Cement Mortars

Panagiota T. Dalla, Ilias K. Tragazikis, Dimitrios A. Exarchos, Konstantinos G. Dassios *, Nektaria M. Barkoula and Theodore E. Matikas

Department of Materials Science and Engineering, University of Ioannina, Dourouti University Campus, GR-45110 Ioannina, Greece; pan.dalla@yahoo.gr (P.T.D.); itragazik@cc.uoi.gr (I.K.T.); dexarch@cc.uoi.gr (D.A.E.); nbarkoul@cc.uoi.gr (N.M.B.); matikas@cc.uoi.gr (T.E.M.)
* Correspondence: kdassios@uoi.gr

Received: 30 December 2018; Accepted: 1 March 2019; Published: 12 March 2019

Abstract: The study investigates the effect of carbon nanotubes on chloride penetration in nano-modified mortars and reports on the physical, electrical, and mechanical performance of the material. Mortars were artificially corroded and their surface electrical surface conductivity as well as flexural and compressive strength were measured. The influence of variable nanotube concentration in accelerated corrosion damage was evaluated. Nanotube concentration was found to significantly affect the permeability of the mortars; improvements in flexural and compressive response of mortars exposed to salt spray fog, compared to virgin specimens, were rationalized upon decreases in the apparent porosity of the materials due to filling of the pores with sodium chloride. Electrical resistivity was found to increase up to two orders of magnitude with respect to the surface value; above the percolation threshold, the property impressively increased up to five orders of magnitude.

Keywords: chloride diffusion; cement mortars; carbon nanotubes; mechanical properties; electrical properties

1. Introduction

Reinforced cement-based structures are expected to remain intact for long periods of time with minimum service requirements. This demands the development and usage of long-lasting building materials. Lately, there is increasing attention towards incorporating nanoparticles in building materials for improvement of their mechanical and electrical performance as well as endowment of multi-functionality to the structure [1–3]. Relevant studies have explored the use, in cement-based materials, of such nano-reinforcements such as nano-$CaCO_3$ [4], nano-SiO_2 [5], as well as carbon-nanofibers (CNFs) [2], graphene nano-platelets [6,7] and carbon-nanotubes (CNTs) [8,9]. It has been reported that homogeneously-dispersed nano-particles can fill inherent voids in cement structures, hence lowering the porosity and increasing their strength and durability [10]. On the other hand, poor design and supervision, unsuitable construction, insufficient materials selection and harsh environments can severely downgrade cement structures [11–13].

One of the major durability issues that the construction industry is facing globally is the materials' structural degradation [14–17]. This may lead to reductions in strength, reinforcement corrosion, as well as aesthetics issues, which would require early repair or replacement of the structure [18]. Corrosion, attacking mainly reinforcements via the pores in the material bulk, is primarily due to carbonation and chloride attack [12,19–22]. Dry environments favor corrosion because of carbonation and carbon oxide excess [23]. On the other hand, localized corrosion in different types of cement-based structures exposed to marine environments may lead to premature structure failure due to chloride ingress through the net of the material pores or via ion incorporation in the mixture [12,24,25].

In concrete, the corrosion process is completed in two stages: In the initial corrosion initiation stage, chlorides penetrate the concrete without causing damage to the material. In the second stage, termed corrosion propagation stage, corrosion elements accumulate until the material's ultimate failure [12]. For concrete structures in marine environments, the service life is the sum of the durations of the two stages, while durability and serviceability of the structures depend greatly on the duration of the initiation stage [12]. A variety of methods are currently available in the literature for prevention of deterioration due to chloride penetration, as are models for prediction of the time of corrosion initiation and total service life [19–24,26]. In order to better assess the phenomenon, knowledge of the different driving forces responsible for the penetration of chloride ions in concrete is indispensable. Possible such driving forces include absorption during wetting and drying cycles [27] and application of hydrostatic pressure, by means of an hydraulic head on the surface of the concrete [18]. However, the most common way in which chloride ions get in contact with concrete reinforcement is diffusion, *i.e.* the movement of chloride ions under a concentration gradient via a continuous liquid phase [28]. The penetration rate depends on the quality of the concrete, its chloride binding capacity, hydration degree, duration of exposure to chlorides, water to cement ratio, temperature and curing time, as well as the presence of supplementary cementing materials [12,18,24,28,29]. A sufficient understanding of these factors and their relationships could provide a good base for the limitation of degradation of cement-based structures.

The degree of resistance of concrete to chloride ion penetration is characterized by the coefficient of diffusion in Fick's corresponding second law [30]. Its calculation requires exposure of the concrete sample to known concentrations of chloride solutions for specific durations followed by measurement of chlorides' concentration at successively larger depths. The values of chloride diffusion coefficient usually vary from 10^{-13} m^2/s to 10^{-10} m^2/s [31] in relation to the concrete properties such as w/cm ratio, the type and proportion of mineral admixtures and cement, the material's compaction and curing state, chloride exposure conditions, as well as the presence of cracks [32]. Currently available experimental methods for the determination of concrete's resistance to chloride ion penetration include the rapid chloride permeability test (RCPT) as shown by Marta Kosior—Kazberuk [33], the rapid chloride test (RCT) method, and the rapid chloride migration test (RMT) [10,34–39]. Probably the most straightforward way of measuring chloride penetration is to immerse the specimens under investigation in salty solutions with different concentrations for specific periods of time depending on the needs of the experiment [40–42]. Ming Jin et al. suggested the measurement of electrical resistivity as a non-destructive method for monitoring chloride ion penetration in concrete structures by means of the electrical properties of graphene-modified cement composites [6].

While considerable research efforts have dealt with concrete behavior with respect to chloride penetration [10,16,38,39,43] and the effect of sodium chloride in mortars [42,44], there is currently a clear lack of knowledge on the effect of carbon nanotube (CNT) presence on chloride penetration in state-of-the-art nano-modified mortars intended for future exploitation in real constructions. The main purpose of the present study is to address the effect of chloride penetration in carbon nanotube-modified cementitious materials subjected to marine environments. For that reason, two different methods, namely the rapid chloride test method (RCT) and the rapid chloride permeability test method (RCPT) were used for the degradation of the specimens, while the mechanical and electrical behavior of the corroded materials was investigated.

2. Experimental Procedures

2.1. Materials and Specimens

Five cement mixtures of varying carbon nanotube concentrations, namely 0.2%, 0.4%, 0.6%, and 0.8% by weight of cement, were prepared according to standard protocol "BS EN 196-1:2005" [45] pertaining to the determination of physical, mechanical, and electrical properties as well as chloride content in cement. A reference mixture without nano-inclusions was also prepared for comparison purposes. All mixtures contained ordinary Portland cement, regular tap water and natural sand. Long multi-wall carbon nanotubes (MWCNT), synthesized via catalytic chemical vapor deposition were used as nano-reinforcements. The tubes, commercially available by Shenzhen Nanotech Port Co. Ltd. (Shenzhen, China), had nominal purity higher than 97% and amorphous carbon content which is less than 3%. Nominal tube diameter ranged from 20 to 40 nm, while their length ranged from 5 to 15 µm. Viscocrete Ultra 300 (Sika AG, Baar, Switzerland), which is a native water-based concrete additive comprised of polycarboxylate polymers, doubled as a superplastisizer and nanotube dispersing agent. Superplastisizer selection was based on its high CNT dispersion efficiency [8,46], excellent resistance to mechanical and chemical attack, as well as its ability to inhibit air entrapment inside the specimens. A second superplastisizer, Viscocrete Ultra 600 (Sika AG, Baar, Switzerland) was added in all fresh mortars, in amounts of approximately 1.5 g, to aid air content stability and workability, hence eliminating contributions of these phenomena to the overall behavior and enabling attribution of findings exclusively to nanotube concentration. In all cases, water to cement ratio was maintained at 0.5.

The experimental procedure for the preparation of nano-reinforced cement mixtures, at tube loadings of 0.2%, 0.4%, 0.6%, and 0.8 wt. % of cement, adopted the three following successive steps. Initially, the superplasticizer and nanotubes were mixed at a ratio of 1.5:1 under magnetic stirring, in regular tap water. Given the fact that for a single mix, BS EN 196-1 requires usage of 1350 g of sand, 450 g of cement, and 225 g of water, the relevant water/CNT ratios were 250 at 0.2% CNT concentration and 125%, 83.5%, and 62.5% at 0.4%, at 0.6%, and 0.8% CNT concentration, respectively. Accordingly, the sand/cement ratio was constant at 3:1. The resultant aqueous suspensions of MWCNTs were ultrasonicated for 90 min at room temperature with a Hielscher UP400S device (Hielscher Ultrasonics GmbH, Teltow, Germany) equipped with a cylindrical 22 mm diameter sonotrode delivering a power throughput of 4500 J/min at a frequency of 24 kHz. The selected combination of ultrasonication duration and energy rate was established as optimum for the suspension homogeneity achievement without tube aspect ratio impairment [47]. The ultrasonicated suspensions were mixed with ordinary Portland cement type "I 42.5N" and natural sand using a laboratory rotary mixer, in low and high speeds sequentially for a total of 4 min, according to standard test method BS EN 196-1:2005 [45]. Immediately after mixing, the workability and air content of the fresh mortars were measured. The remaining fresh mixture was poured into oiled metallic molds with internal dimensions of $40 \times 40 \times 160$ mm for flexural and compressive test specimen production and $150 \times 150 \times 150$ mm for chloride penetration and electrical conductivity measurement specimen, where they were left for 24 h before demolding. The specimens were subsequently placed into a 100% humidity room for 28 days before being transferred into a laboratory corrosive environment.

2.2. Properties in Fresh Conditions

Fresh mortar consistency was determined following flow table tests, carried out according to European Standard BS EN 1015-3:1999 [48]. Consistency is a measure of the fluidity and/or wetness of the fresh mortar and is indicative of the deformability of the fresh mortar when subjected to certain types of stresses [48]. The mortar was introduced into the mold carefully in two layers while excess mortar was removed. Subsequently, the mold was slowly raised, and the mortar was spread out on the flow table disc by jolting the table 15 times, approximately one per second. This causes the mortar to spread further, in a roughly circular shape. The flow diameter is the average of the maximum diameter of the pool of fresh mortar and the diameter at right angles. This average diameter value constitute the flowability of the mixture after subtraction of the mold diameter, 100 mm [48]. Mixture air content was determined after European Standard BS EN 1015-7:1999 [49]. Therein, using an air entrainment meter for mortars (TESTING Bluhm & Feuerherdt GmbH—Berlin, Germany), filling capacity of 1 L, water was introduced on top of the mortar surface and forced into the mortar by means of applied air-pressure hence displacing air from within the pores. The corresponding drop in water level reflects the volume of air displaced from the mortar with the direct reading of the air content in percentage.

2.3. Chloride Penetration

The resistance to chloride penetration of the modified mortars with different CNT loadings was qualitatively evaluated using two effective methods, namely RCPT and RCT, suitable for concrete characterization. The specimens used in these tests were the cores of $150 \times 150 \times 150$ mm^3 cubic samples.

2.3.1. RCPT Method

Rapid chloride ion penetrability test (RCPT) was conducted according to ASTM C1202-97 [50] using the PROOVE'it$^©$ system (Germann Instruments—Copenhagen, Denmark). It monitors the amount of electrical current passed through a 50 mm thick, 100 mm in diameter specimen, which was positioned in a measuring cell and supplied with a direct current (DC) electric current of 60 V for 6 h; the relevant apparatus is illustrated in Figure 1. In each side of the specimen rested a fluid reservoir. One reservoir was filled with 3% NaCl and was connected to the negative terminal, while the other was filled with 0.3 mol/dm^3 solution of NaOH and was connected to the positive terminal of the power supply.

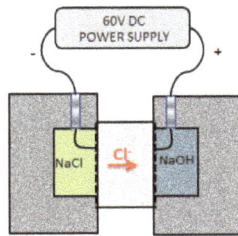

Figure 1. Experimental set up for the rapid chloride permeability test method.

The total charge, in Cb, passing through the specimen is representative of its resistance to chloride ion penetration. The chloride ion penetrability is evaluated qualitatively according to Table 1.

Table 1. Chloride ion penetrability based on charge passed.

Charge Passed [Cb]	Chloride Ion Penetrability
>4000	high
2000 ÷ 4000	moderate
1000 ÷ 2000	low
100 ÷ 1000	very low
<100	negligible

During the test, the solution temperature was monitored and maintained in the range of 20 to 25 °C, as higher temperatures aid acceleration of chloride ion transport, hence resulting in non-representative data.

2.3.2. RCT Method

Evaluation of the chloride penetration features of nano-modified mortars was also carried out by exposing specimens to salt spray environments according to ASTM B117 [51] specifications. As laboratory conditions are much more aggressive than natural ones, to accelerate the phenomenon, severe corrosion states in short times were anticipated. Compared to other laboratory-grade accelerated corrosion tests, the salt spray test was considered more representative of the natural coastal environment [52]. Therein, after removal from the 100% humidity room, nano-modified mortars were surface dried and then naturally dried in the laboratory environment. Subsequently, they were sealed with insulating exterior paint on four sides, including the top side of the mold and two opposite open sides permitted chloride diffusion through a predefined path. Coated samples were placed in a VSC 450 salt spray corrosion test chamber (Vötsch Industrietechnik, Balingen, Germany) (Figure 2a) for 100 days, with a sealed side resting at the bottom of the chamber (Figure 2b).

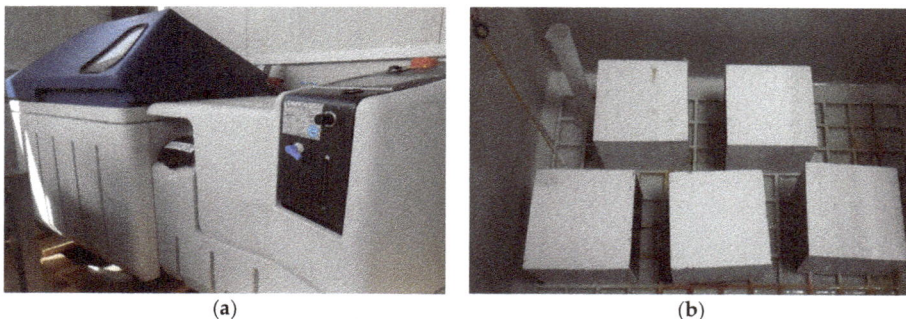

(a)

(b)

Figure 2. (**a**) Salt spray chamber and (**b**) sealed specimens inside the chamber.

Salt solution prepared by dissolving five parts, by mass, of sodium chloride (NaCl) into 95 parts of deionized water, was introduced into the chamber's solution reservoir. The utilized salt contained less than 0.3 percent impurities to avoid impurities acting as corrosion inhibitors. Salt spray solution acidity, measured at 25 °C, was maintained in the pH range from 6.5 to 7.2, while the temperature inside the chamber was maintained at 35 °C [52]. After the end of the 100-day exposure period, nano-reinforced mortars were carefully removed from the chamber and gently washed with clean tap water to remove any surface salt deposit before being naturally dried.

The dried mortars were prepared for testing by the following methodology. Initially, a cylindrical core of the exposed sides to the NaCl was drilled out of each cubic specimen. The diameter of the cores was maintained within 95 mm to 105 mm while sectioning of the total length provided cylindrical samples of an approximate height of about 65 mm, such as the ones shown in Figure 3.

Figure 3. Testing cores drilled off from mortars exposed to accelerated corrosion.

Subsequently, 1.5 g of powder ground from incrementally increasing depths inside the nano-modified material were removed using a profile grinder (Germann Instruments—Copenhagen, Denmark) and collected for chloride ion measurement. The grinding depth increment was kept as low as 1 mm, to allow for high accuracy measurement; 60 successive depths were interrogated. Next, ground powders originating from mortars with different CNT content were tested separately according to the rapid chloride test (RCT) method. In this way, the amount of chloride calculated using the RCT electrode, which was initially calibrated, was related to CNT loading. Profiles of chloride content versus the depth below the exposed surface were determined.

Assuming that diffusion is a one dimensional process, diffusion of chloride ions into the materials is governed by Fick's second law of diffusion (steady state diffusion) as follows:

$$\frac{\partial C(x)}{\partial t} = D\frac{\partial^2 C}{\partial x^2} \tag{1}$$

where C represents chloride ion concentration, x is the depth from the surface of concrete, t represents time, and D is the diffusion coefficient.

Solving Equation (1) requires knowledge of initial and boundary conditions as well as the value of the diffusion coefficient, D. For a constant diffusion coefficient and under the assumptions that chloride ion concentration on the material's surface is constant (Cs = C(0,t)), the initial concentration (Ci) is zero and the concentration at an infinite point, quite deep in relation to the surface, is also zero. Equation (1) converts to Crank's solution [53] as follows:

$$C(x,t) = Cs - (Cs - Ci)\text{erf}\left(\frac{x}{2\sqrt{Dt}}\right) \tag{2}$$

where C(x,t) indicates the chloride ion concentration at a depth of x, measured from the surface in mm, at an elapsed time t, measured in years, from the start of the chloride exposure

Cs is the chloride concentration (% of materials mass) at the material's surface

Ci is the initial or background chloride concentration (% of materials mass)

D represents the chloride diffusion coefficient (mm^2/year)

Equation (2) describes the variation of chloride ion content as a function of the distance x from the surface of the sample, after an elapsed time t since initial exposure to a constant surface chloride concentration of Cs. The values of parameters Cs and D are determined using the least-squares curve fitting that permits general non-linear regression analysis.

2.4. Porosimetry

The pore structure of the mortars modified with variable nanotube concentrations was analyzed by mercury porosimetry in a PoreMaster 60 porosimeter (Quantachrome GmbH & Co. KG, Odelzhausen, Germany). The measurements were performed on small nano-modified mortar volumes, less than 1 cm^3, extracted from specimens after mechanical testing failure.

2.5. Electrical Conductivity Measurements

Surface electrical conductivity was incrementally measured at the exposed mortar surface, after step grinding of 10 mm of material. DC electrical conductivity was measured using a custom-built contact electrical conductivity probe connected to an ultra-high precision digital electrometer/high resistance meter (Keithley 6517B, Tektronix Inc., Beaverton, OR, USA) capable of measuring resistances up to 1018 with a 10×10^{-18} A current resolution. The probe, presented in detail in reference [2], consists of a circular head comprised of 22 concentrically-arranged spring-loaded pin electrodes with conductive rubber ends for optimal contact with the non-planar cement surfaces. The head rested on a z-translational stage, which could be lowered and brought into contact with the specimen at constant force by means of a lever. Conductivity, assumed exclusively attributable to nanotube presence, is anticipated to vary with interrogation depth as a result of variable chloride ion concentration at the instant depth.

2.6. Mechanical Performance

Mechanical characterization of corroded nano-modified mortars under the four point bending test configuration was performed on $40 \times 40 \times 160$ mm^3 specimens on an Instron 5967 testing frame (Instron, Norwood, MA, USA) equipped with a 30 kN loadcell according to standard test protocol ASTM C1609 [54]. Prism halves occurring after the catastrophic bending test, were subjected to compression testing, at a load rate of 2400 N/s, which corresponds to a stress rate of 1.5 MPa/s for the 40x40 mm^2 tested area, following the standard protocol EN 196-1:2005 [41].

3. Results and Discussion

3.1. Workability and Air Content of Fresh Mortars

Fresh mortar properties are the ones which determine the application range of the material. During the course of the present study, workability as well as air content of the fresh nano-reinforced mortars was maintained, in similar values throughout mixtures of variable nanotube concentration, by addition of differential amounts of Viscocrete 600 superplasticizer as needed. As can be observed in Figure 4, the workability of the mortars reinforced with different concentrations of carbon nanotubes does not vary significantly from the starting, plain-mortar value. It must be recalled that the sole dispersion assistive agent used in the mixture, which affects workability, was Viscocrete Ultra 300, a native concrete additive which has proven highly effective for CNT dispersion, hence rendering surfactant usage and chemical tube functionalization, unnecessary [8,55]. The invariance of workability with respect to tube loading is a favorable finding, it ensures that the latter can act as the sole parameter responsible for any documented variation in properties.

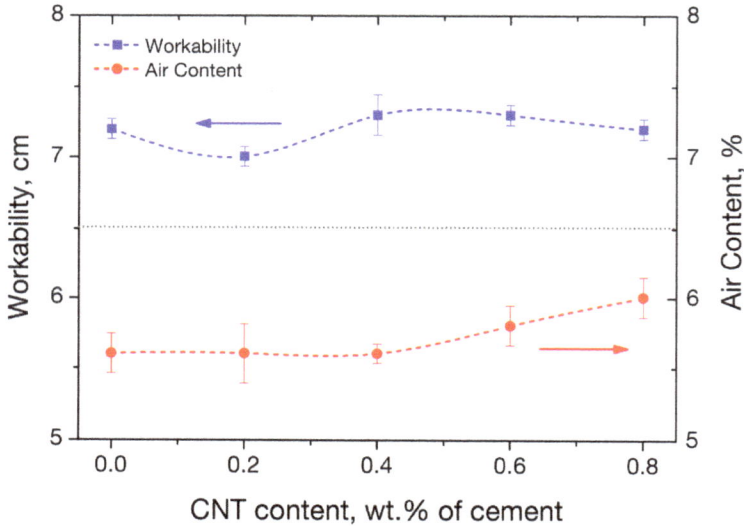

Figure 4. Change of flowability and air content across nano-modified mortars with different concentrations of carbon nanotubes.

Following completion of the flow table test, proper volumes of fresh mortars were used for air content testing. Results from each mixture are plotted as a function of the tube content in Figure 4. Here again, all specimens are observed to share similar air contents, a finding which confirms the workability results as well as the role of Viscocrete 600 superplastisizer in the mixing procedure.

3.2. Chloride Profile of the CNTs Nano-Modified Mortars

3.2.1. RCPT Method

The resistance to chloride penetration was initially evaluated qualitatively using the RCPT method; results of current passing through nano-modified mortars with different CNT loadings after curing duration of 28 days are illustrated in Figure 5. It is therein observed that throughout the whole test duration, current was lower for plain specimens and increased with CNT concentration. The current passing through the sample with 0.8% wt. CNTs was approximately 20% higher than in the sample with no nano-reinforcement. The observed behavior is rationalized upon the anticipated increase in conductivity endowed by higher concentrations of the conductive filler and compares favorably to relevant recent findings [56,57]

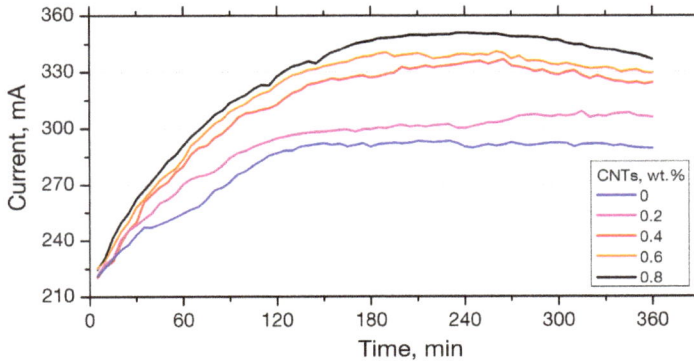

Figure 5. Current passing through nano-modified mortars with different amount of carbon nanotubes (CNTs).

The electric charge allowed to pass through the nano-modified mortars is illustrated in Figure 6 as a function of tube concentration. Therein, negligible charge variation is observed with CNT loading, while according to Table 1 classification, all samples appear to exhibit high chloride ion penetrability. Furthermore, it is shown that the charge passing through the plain specimens is almost identical to that passing through nano-modified mortars, a finding which suggests complete invariance of the particular behavior to carbon nanotube presence. In view of these findings, three factors in the RCPT method must not be neglected: (i) That the current passed is related to all ions in the pore solution, not just chloride ions, (ii) that measurements are taken before steady-state migration is achieved, and (iii) that the high voltage applied may lead to temperature increase which, in turn, may influence the measurement [18,28].

3.2.2. RCT Method

As aforementioned, following 100 day exposure to salt spray, profile grinding at an increment of 1 mm, a total of 60 interrogations were performed. Figure 7 plots the typical chloride concentration profiles, expressed in % mass of mortar as a function of interrogation depth.

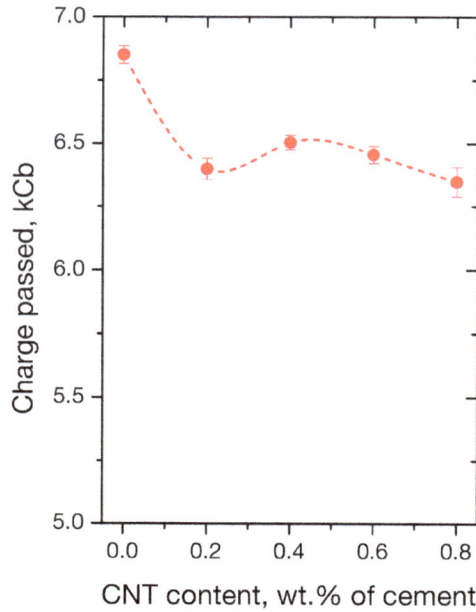

Figure 6. Total charge passed through specimens with different CNT loadings.

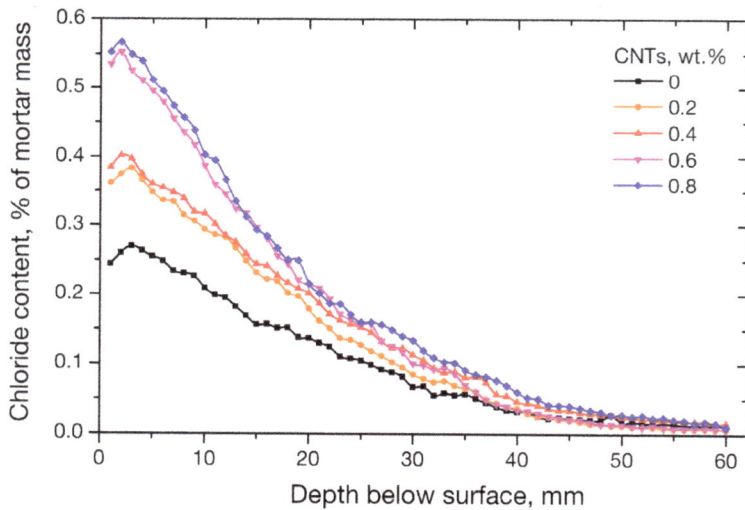

Figure 7. Chloride profile of the nano-modified mortars at different depths.

It is observed that the profiles had an almost identical shape throughout mortars of variable CNT concentration, with chloride content being at a maximum near the surface and decreasing with interrogation depth. For all specimens, the first one or two measurements, corresponding to the first two grinding steps, show less chloride ions than expected as a result of the aforementioned washout, which was performed following the exposure procedure [58]. These initial data points were not considered in the least-squares regression analysis for the obtainment of the best-fit values of the parameters of Equation (2). Moreover, it is observed that after the first 50 mm below the exposed surface, ion content drops to less than 0.05% where nanotube presence does not make a difference to the total recorded behavior. Most importantly, at all depths, chloride content was consistently higher for samples with higher CNT concentrations, a finding which signifies that chloride ion concentration is directly affected by nanotube presence. According to Figure 7, the influence is moderate for CNT loadings of 0.2% and 0.4% wt. of cement and more intense in mortars loaded with 0.6% and 0.8% nanotubes. Chloride concentration in mortars with 0.8% wt. cement CNTs is almost 50% higher than that of plain mortars. It is believed that after drying the mortars that were exposed to salt spray, chloride ions are still present in the cement matrix, although the network of pore water which initially carried them and acted as an electrical pathway, is now removed. Resultant chloride-containing crystals are electrically conductive and are now precipitated between networks of conducting carbon nanotubes. This facilitates electrical current flow through the material and absorption of a considerable amount of chloride ions. Inarguably, bigger amounts of chloride ions manage to travel effectively and are trapped to the inner part of the structure as CNT concentration increases. The chloride diffusion coefficient values calculated for nano-modified mortars exposed to 5% NaCl solution following the analysis presented in Section 2.3.2 are shown in Figure 8. It is therein observed that the coefficient increases for the first two nanotube loadings and is at a maximum at a CNT concentration of 0.4% wt. of cement above which it drops again by approximately 30% of the maximum value, to the level of the plain specimens or slightly lower. Joining information from Figures 7 and 8, it can be argued that higher chloride diffusion coefficient values do not necessarily signify larger amounts of chlorides penetration in the specimen. This is because the coefficient expresses how fast the chlorides are transported in the material and depends on external factors such as the degree of hydration and saturation, as well as porosity.

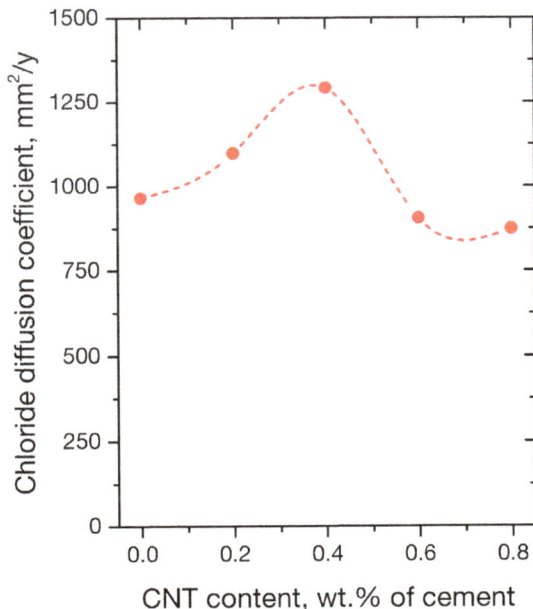

Figure 8. Chloride diffusion coefficient for each nano-modified mortar.

3.3. Porosity

The pore size distributions measured by mercury porosimetry in mortars with variable carbon nanotube contents are presented in Figure 9. It is therein observed that all mortars exhibited overall comparable pore distribution curves with the main features being (i) the dominant peak at ca.

6×10^{-3} µm, on 6 nm, which is fixed throughout variable nano-inclusion concentrations and is inarguably attributed to intra-nanotube porosity, i.e., the over volume corresponding to the internal open space of the cylindrical multi-walled tubes, and (ii) a pore range of non-consistent peaks up to 0.2 µm.

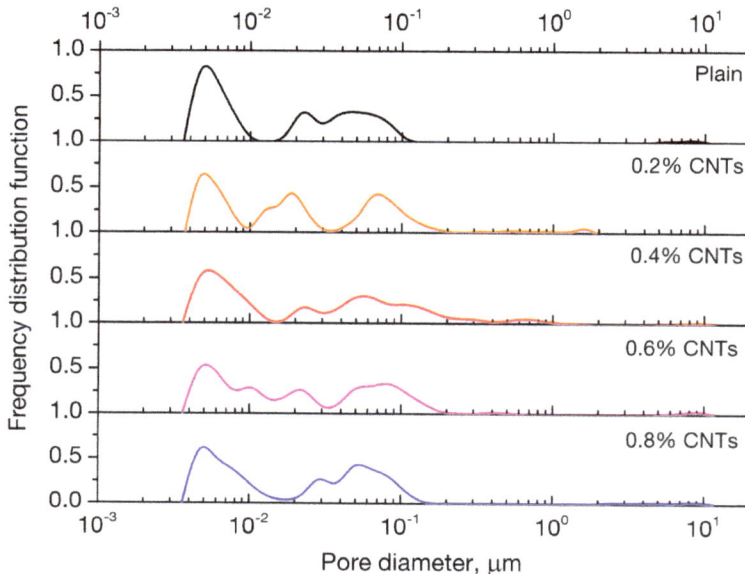

Figure 9. Pore size distributions of nano-modified mortars.

The actual porosity values for each mortar with variable CNT concentration are given in Table 2. It is observed that the total porosity drops with an increasing CNT concentration up to 0.4 wt. %, above which it increases to values even higher than the plain case. This may be due to nanotube entanglement issues at higher nano-inclusion concentrations.

Table 2. Porosity values for nano-modified mortars.

CNT Content % wt. of Cement	Porosity (%)
0	16.22
0.2	14.64
0.4	12.68
0.6	19.28
0.8	20.57

Additionally, chloride diffusion coefficients of corroded mortars are plotted as a function of respective mortar porosity in Figure 10. An inverse correlation between the two parameters is therein observed, with diffusion coefficients values decreasing with total porosity. This finding can be rationalized upon considering that more porous microstructures can block the rate of chloride penetration inside the testing mortars.

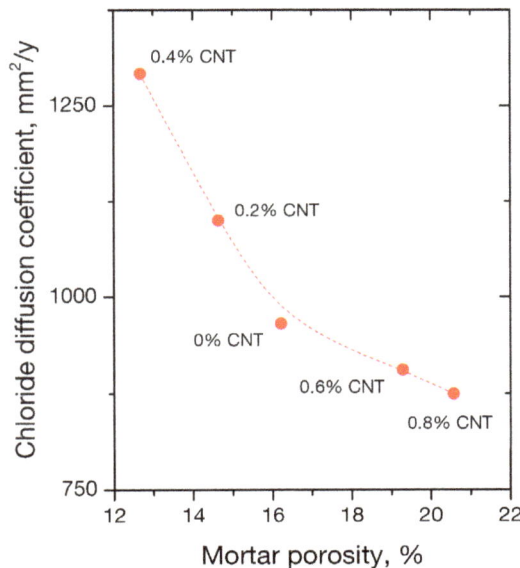

Figure 10. Chloride diffusion coefficient a function of mortar porosity.

Based on the above results, chloride diffusion in the processed mortars appears to be strongly linked with CNT concentration. This is due to a combination of mechanisms, one related to the raise of electric conductivity for CNT concentrations above the percolation threshold [2], the other associated to the fact that CNT addition lowers porosity values in the material. Hence, CNTs affect chloride diffusivity both directly, by affecting electrical transport properties, but also indirectly, by affecting the microstructure.

3.4. Electrical Properties of Nano-Modified Mortars

The effect of salt spray (fog) corrosion test on the mortars' electrical properties was investigated by measurement of surface electrical resistivity at successive depths inside the ground mortars, with a step of 10 mm, as shown in Figure 11. The depicted changes in electrical resistivity at successive depths below the exposed surface are indicative of physical and chemical changes taking place into the material [33]. Specimens with 0%, 0.2% and 0.4% nano-inclusion concentrations, demonstrate a gradual increase in surface electrical resistivity with interrogation depth and reach, at a depth of 60 mm, a practically common value. This indicates the absence of chloride ions at such depths, independently of the concentration of nano-inclusions in the compositions. On the other hand, resistivity in samples with 0.6% and 0.8% nano-inclusion concentrations commences at the surface with significantly lower values than in the previous case, and follows a gradually increasing trend with depth, until it catches up with the behavior of the lower CNT concentrations. This may be attributed to the electrical percolation threshold of the nano-modified mortars wherein a conductive filler concentration of 0.6% wt. cement CNTs is already sufficient for the creation of a conductive electrical path [2]. Electrical resistivity appears to increase with interrogation depth below the exposed surface, which indicates that chloride concentration decreases with depth. On the contrary, porosity is one of the main parameters affecting the electrical resistivity in a cement-based material. The lower porosity values of specimens with 0.6% and 0.8% wt. to cement carbon nanotubes (Table 2), may indicate that the electrical current takes the path of least resistance inside the specific structures.

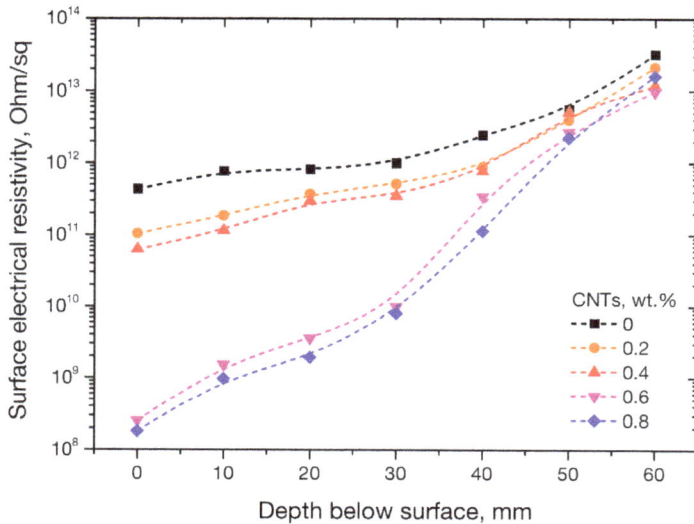

Figure 11. Surface electrical resistivity with respect to the depth below the exposed surface for all the nano-modified cement mortar.

3.5. Mechanical Performance of Nano-Modified Mortars

The effect of carbon nanotube concentration on the flexural and compressive strengths of the nano-modified mortars, both before and after exposure to salt spray conditions for a duration of 100 days, is presented in Figure 12. Virgin samples exhibited negligible fluctuation of the compressive strength with respect to CNT content, while flexural strength showed a small increase up to 0.4% CNT loadings. These findings compare favorably with previous research dedicated to the mechanical properties of CNT-modified mortars [8]. It is observed that exposure to sea fog resulted in improved flexural and mechanical properties for all specimens, irrespective of nanotube concentration. Concerning flexural strength, the property appears to increase by approximately 10%, commonly for mortars with CNT concentrations up to 0.4 wt. %, as a result of exposure to NaCl. The highest increase with respect to unexposed specimens is 22% and is noted for mortars with 0.8 wt. % nanotube content. These results are not incompatible with the porosimetry findings, wherein the inner pores of the specimens are filled with salt from the water solution that mortars are subjected to. Because of the lower porosity of the specimens with 0.6% and 0.8%, the percent increase of flexural strength of the specimens up to 0.4% nanotube concentration is greater. On the other hand, compressive strength appears to increase after the exposure to salt spray conditions for all the investigated ranges of carbon nanotube loadings and is generally constant at about 10%.

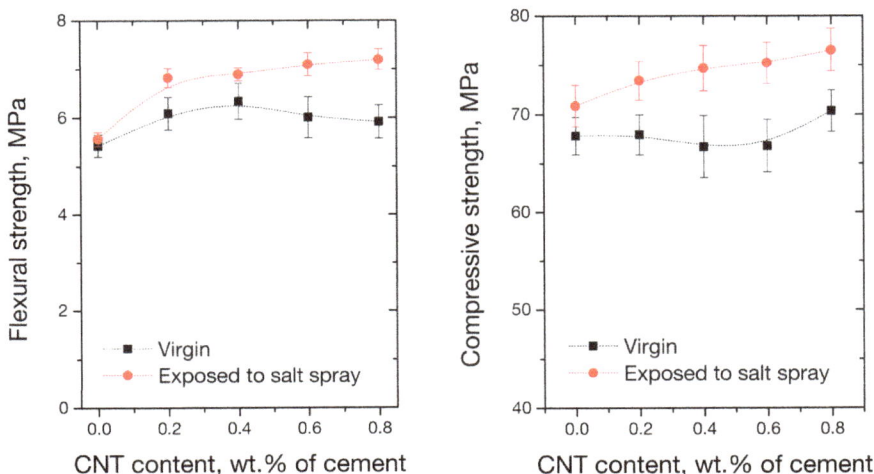

Figure 12. Mechanical properties of the nano-modified mortars before and after the exposure to the salt spray chamber.

4. Conclusions

The effect of carbon nanotube concentration on chloride penetration in nano-modified mortars was investigated, and their physical, electrical, and mechanical performance were assessed. Mortar workability and air content in fresh conditions were kept constant among variable nanofiller concentrations, thus as to not influence the ion penetration inside the nano-modified specimens, and allow for rationalization of the findings by means of variable nanotube loading, exclusively. The main conclusions can be summarized as follows:

1. Using the rapid chloride permeability test method (RCPT) for plain and nano-reinforced mortars, all mortars were observed to exhibit high ion permeability. It must not be neglected that the RCPT-measurable permeability reflects not only chloride ions, but the total of ions contained to the mixture.
2. Nano-inclusions were found to affect the permeability of the mortars since the absorbed ion chlorides increased with nanotube concentration. Diffusion coefficient (Da) did not follow the same trend; it increased up to 0.4 wt. % of nanofiller content and decreased for higher contents of 0.6% and 0.8%. Porosimetry findings are compatible with the observed behavior.
3. The flexural and compressive properties of mortars exposed to salt spray fog present improved values compared to the virgin specimens. The observed improvement was rationalized upon sodium chloride filling the pores, which resulted in a decrease in apparent porosity and an increase in material strength.
4. Electrical resistivity, measured incrementally every 10 mm below the surface of the specimen, presented notable variations as a result of the specimens' subjection to NaCl solution. For mortars with 0, 0.2, and 0.4 wt. % CNTs, resistivity increased up to two orders of magnitude with interrogation depth. On the other hand, mortars with percolated conductive nanofiller networks at CNT concentrations of 0.6 and 0.8 wt. % CNTs, exhibited increases in electrical resistivity up to five orders of magnitude with respect to the surface value.

Author Contributions: Conceptualization, K.G.D. and T.E.M.; methodology, K.G.D., N.M.B., and T.E.M.; validation, P.T.D., I.K.T., and D.A.E.; formal analysis, P.T.D., I.K.T., D.A.E. and N.M.B.; investigation, P.T.D., I.K.T., D.A.E., and N.M.B.; data curation, P.T.D., I.K.T., D.A.E. and N.M.B.; writing—original draft preparation, P.T.D. and K.G.D.; writing—review and editing, P.T.D. and K.G.D.; visualization, X.X.; supervision, K.G.D., N.M.B. and T.E.M.; project administration, K.G.D. and T.E.M.

Funding: This research received no external funding.

Conflicts of Interest: The authors declare no conflict of interest.

References

1. Sanchez, F.; Sobolev, K. Nanotechnology in concrete—A review. *Constr. Build. Mater.* **2010**, *24*, 2060–2071. [CrossRef]
2. Dalla, P.T.; Dassios, K.G.; Tragazikis, I.K.; Exarchos, D.A.; Matikas, T.E. Carbon nanotubes and nanofibers as strain and damage sensors for smart cement. *Mater. Today Commun.* **2016**, *8*, 196–204. [CrossRef]
3. Tragazikis, I.; Dalla, P.; Exarchos, D.; Dassios, K.; Matikas, T. Nondestructive evaluation of the mechanical behavior of cement based nanocomposites under bending. *Proc. SPIE* **2015**, *9436*, 94360F.
4. Kawashima, S.; Seo, J.-W.T.; Corr, D.; Hersam, M.C.; Shah, S.P. Dispersion of CaCO3 nanoparticles by sonication and surfactant treatment for application in fly ash–cement systems. *Mater. Struct.* **2014**, *47*, 1011–1023. [CrossRef]
5. Hou, P.; Qian, J.; Cheng, X.; Shah, S.P. Effects of the pozzolanic reactivity of nanoSiO2 on cement-based materials. *Cem. Concr. Compos.* **2015**, *55*, 250–258. [CrossRef]
6. Jin, M.; Jiang, L.; Lu, M.; Bai, S. Monitoring chloride ion penetration in concrete structure based on the conductivity of graphene/cement composite. *Constr. Build. Mater.* **2017**, *136*, 394–404. [CrossRef]
7. Tragazikis, I.K.; Dassios, K.G.; Dalla, P.T.; Exarchos, D.A.; Matikas, T.E. Acoustic emission investigation of the effect of graphene on the fracture behavior of cement mortars. *Eng. Fract. Mech.* **2018**, in press. [CrossRef]
8. Tragazikis, I.K.; Dassios, K.G.; Exarchos, D.A.; Dalla, P.T.; Matikas, T.E. Acoustic emission investigation of the mechanical performance of carbon nanotube-modified cement-based mortars. *Constr. Build. Mater.* **2016**, *122*, 518–524. [CrossRef]
9. Dheeraj Swamy, B.L.P.; Raghavan, V.; Srinivas, K.; Narasinga Rao, K.; Lakshmanan, M.; Jayanarayanan, K.; Mini, K.M. Influence of silica based carbon nano tube composites in concrete. *Adv. Compos. Lett.* **2017**, *26*, 12–17.
10. Du, H.; Gao, H.J.; Pang, S.D. Improvement in concrete resistance against water and chloride ingress by adding graphene nanoplatelet. *Cem. Concr. Res.* **2016**, *83*, 114–123. [CrossRef]
11. Darmawan, M.S.; Bayuaji, R.; Husin, N.A.; Anugraha, R.B. Case Study of Remaining Service Life Assessment of a Cooling Water Intake Concrete Structure in Indonesia. *Adv. Civ. Eng.* **2014**, *2014*, 16. [CrossRef]
12. Hodhod, O.A.; Ahmed, H.I. Modeling the service life of slag concrete exposed to chlorides. *Ain Shams Eng. J.* **2014**, *5*, 49–54. [CrossRef]
13. Neville, A. Chloride attack of reinforced concrete: An overview. *Mater. Struct.* **1995**, *28*, 63. [CrossRef]
14. Zhu, J.; Zhang, Y.; Zhao, D. Durability Assessment of an RC Railway Bridge Pier under a Chloride-Induced Corrosion Environment. In Proceedings of the Fifth International Conference on Transportation Engineering, Dalian, China, 26–27 September 2015.
15. Zeng, L.; Song, R. Controlling chloride ions diffusion in concrete. *Sci. Rep.* **2013**, *3*, 3359. [CrossRef]
16. Song, Z.; Jiang, L.; Liu, J.; Liu, J. Influence of cation type on diffusion behavior of chloride ions in concrete. *Constr. Build. Mater.* **2015**, *99*, 150–158. [CrossRef]
17. Otieno, M.; Beushausen, H.; Alexander, M. Chloride-induced corrosion of steel in cracked concrete – Part I: Experimental studies under accelerated and natural marine environments. *Cem. Concr. Res.* **2016**, *79*, 373–385. [CrossRef]
18. Stanish, K.D.; Hooton, R.D.; Thomas, M.D. *Testing the Chloride Penetration Resistance of Concrete: A Literature Review*; FHWA Contract DTFH61; Department of Civil Engineering, University of Toronto: Toronto, ON, Canada, 1997.
19. Liang, M.T.; Wang, K.L.; Liang, C.H. Service life prediction of reinforced concrete structures. *Cem. Concr. Res.* **1999**, *29*, 1411–1418. [CrossRef]
20. Liang, M.T.; Hong, C.L.; Liang, C.H. Service life prediction of existing reinforced concrete structures under carbonation-induced corrosion. *J. Chin. Inst. Civ. Hydraul. Eng.* **1999**, *11*, 485–492.
21. Morinaga, S. *Prediction of Service Lives of Reinforced Concrete Buildings Based on Rate of Corrosion of Reinforcing Steel*; Report No. 23; Shimizu Corp: Tokyo, Japan, 1988; pp. 82–89.
22. Jamali, A.; Angst, U.; Adey, B.; Elsener, B. Modeling of corrosion-induced concrete cover cracking: A critical analysis. *Constr. Build. Mater.* **2013**, *42*, 225–237. [CrossRef]

23. Ranjith, A.; Rao, K.B.; Manjunath, K. Evaluating the effect of corrosion on service life prediction of RC structures—A parametric study. *Int. J. Sustain. Built Environ.* **2016**, *5*, 587–603. [CrossRef]

24. Roa-Rodriguez, G.; Aperador, W.; Delgado, A. Calculation of Chloride Penetration Profile in Concrete Structures. *Int. J. Electrochem. Sci.* **2013**, *8*, 5022–5035.

25. Kim, H.K.; Jang, J.G.; Choi, Y.C.; Lee, H.K. Improved chloride resistance of high-strength concrete amended with coal bottom ash for internal curing. *Constr. Build. Mater.* **2014**, *71*, 334–343. [CrossRef]

26. Liang, M.T.; Huang, R.; Feng, S.A.; Yeh, C.J. Service life prediction of pier for the existing reinforced concrete bridges in chloride-laden environment. *J. Mar. Sci. Technol.* **2009**, *17*, 312–319.

27. Gergely, J.; Bledsoe, J.E.; Tempest, B.Q.; Szabo, I.F. *Concrete Diffusion Coefficients and Existing Chloride Exposure in North Carolina*; Project No. HWY-2004-12; Department of Civil Engineering, University of North Carolina at Charlotte: Charlotte, NC, USA, 2006.

28. Altaf Ahmad, A.K. Chloride ion migration/diffusion through concrete and test methods. *Int. J. Adv. Sci. Tech. Res.* **2013**, *6*, 151–180.

29. Angst, U.; Elsener, B.; Larsen, C.K.; Vennesland, Ø. Critical chloride content in reinforced concrete—A review. *Cem. Concr. Res.* **2009**, *39*, 1122–1138. [CrossRef]

30. Paul, A.; Laurila, T.; Vuorinen, V.; Divinski, S.V. Fick's Laws of Diffusion. In *Thermodynamics, Diffusion and the Kirkendall Effect in Solids*; Springer International Publishing: Cham, Switzerland, 2014; pp. 115–139.

31. Bertolini, L.; Elsener, B.; Pedeferri, P.; Polder, R. *Corrosion of Steel in Concrete: Prevention, Diagnosis, Repair*; Wiley-VCH: Weinheim, Germany, 2004.

32. Shi, X.; Yang, Z.; Liu, Y.; Cross, D. Strength and corrosion properties of Portland cement mortar and concrete with mineral admixtures. *Constr. Build. Mater.* **2011**, *25*, 3245–3256. [CrossRef]

33. Kosior-Kazberuk, M.; Jezierski, W. Evaluation of concrete resistance to chloride ions penetration by means of electric resistivity monitoring. *J. Civ. Eng. Manag.* **2005**, *11*, 109–114. [CrossRef]

34. Liu, J.; Wang, X.; Qiu, Q.; Ou, G.; Xing, F. Understanding the effect of curing age on the chloride resistance of fly ash blended concrete by rapid chloride migration test. *Mater. Chem. Phys.* **2017**, *196*, 315–323. [CrossRef]

35. Dodds, W.; Christodoulou, C.; Goodier, C.; Austin, S.; Dunne, D. Durability performance of sustainable structural concrete: Effect of coarse crushed concrete aggregate on rapid chloride migration and accelerated corrosion. *Constr. Build. Mater.* **2017**, *155*, 511–521. [CrossRef]

36. Ying, J.; Zhou, B.; Xiao, J. Pore structure and chloride diffusivity of recycled aggregate concrete with nano-SiO2 and nano-TiO2. *Constr. Build. Mater.* **2017**, *150*, 49–55. [CrossRef]

37. Snyder, K.A.; Ferraris, C.; Martys, N.S.; Garboczi, E.J. Using Impedance Spectroscopy to Assess the Viability of the Rapid Chloride Test for Determining Concrete Conductivity. *J. Res. Natl. Inst. Stand. Technol.* **2000**, *105*, 497–509. [CrossRef]

38. Valipour, M.; Shekarchi, M.; Arezoumandi, M. Chlorine diffusion resistivity of sustainable green concrete in harsh marine environments. *J. Clean. Prod.* **2017**, *142*, 4092–4100. [CrossRef]

39. Madani, H.; Bagheri, A.; Parhizkar, T.; Raisghasemi, A. Chloride penetration and electrical resistivity of concretes containing nanosilica hydrosols with different specific surface areas. *Cem. Concr. Compos.* **2014**, *53*, 18–24. [CrossRef]

40. Song, Z.; Jiang, L.; Zhang, Z.; Xiong, C. Distance-associated chloride binding capacity of cement paste subjected to natural diffusion. *Constr. Build. Mater.* **2016**, *112*, 925–932. [CrossRef]

41. Arya, C.; Buenfeld, N.R.; Newman, J.B. Factors influencing chloride-binding in concrete. *Cem. Concr. Res.* **1990**, *20*, 291–300. [CrossRef]

42. Kim, H.-K. Chloride penetration monitoring in reinforced concrete structure using carbon nanotube/cement composite. *Constr. Build. Mater.* **2015**, *96*, 29–36. [CrossRef]

43. Aldea, C.-M.; Young, F.; Wang, K.; Shah, S.P. Effects of curing conditions on properties of concrete using slag replacement. *Cem. Concr. Res.* **2000**, *30*, 465–472. [CrossRef]

44. Elfmarkova, V.; Spiesz, P.; Brouwers, H.J.H. Determination of the chloride diffusion coefficient in blended cement mortars. *Cem. Concr. Res.* **2015**, *78*, 190–199. [CrossRef]

45. Dalton, R. US researchers fear job losses from privatization drive. *Nature* **2003**, *424*, 478. [CrossRef]

46. Alafogianni, P.; Dassios, K.; Tsakiroglou, C.D.; Matikas, T.E.; Barkoula, N.M. Effect of CNT addition and dispersive agents on the transport properties and microstructure of cement mortars. *Constr. Build. Mater.* **2019**, *197*, 251–261. [CrossRef]

47. Dassios, K.G.; Alafogianni, P.; Antiohos, S.K.; Leptokaridis, C.; Barkoula, N.-M.; Matikas, T.E. Optimization of sonication parameters for homogeneous surfactant-assisted dispersion of multiwalled carbon nanotubes in aqueous solutions. *J. Phys. Chem. C* **2015**, *119*, 7506–7516. [CrossRef]

48. Dalton, R. Museum trustees quit after row over sale of artefacts. *Nature* **2003**, *424*, 360. [CrossRef] [PubMed]

49. Dalton, R. Ants join online colony to boost conservation efforts. *Nature* **2003**, *424*, 242. [CrossRef] [PubMed]

50. *ASTM C1202-97—Standard Test Method for Electrical Indication of Concrete's Ability to Resist Chloride Ion Penetration*; ASTM International: West Conshohocken, PA, USA, 1997.

51. *ASTM B117-16—Standard Practice for Operating Salt Spray (Fog) Apparatus*; ASTM International: West Conshohocken, PA, USA, 2016.

52. Papadopoulos, M.P.; Apostolopoulos, C.A.; Alexopoulos, N.D.; Pantelakis, S.G. Effect of salt spray corrosion exposure on the mechanical performance of different technical class reinforcing steel bars. *Mater. Des.* **2007**, *28*, 2318–2328. [CrossRef]

53. Crank, J. *The Mathematics of Diffusion*, 2nd ed.; Clarendon Press: Oxford, UK, 1975.

54. *ASTM C1609/C1609M-05—Standard Test Method for Flexural Performance of Fiber-Reinforced Concrete (Using Beam with Third-Point Loading)*; ASTM International: West Conshohocken, PA, USA, 2005.

55. Alafogianni, P.; Dassios, K.; Farmaki, S.; Antiohos, S.K.; Matikas, T.E.; Barkoula, N.M. On the efficiency of UV–vis spectroscopy in assessing the dispersion quality in sonicated aqueous suspensions of carbon nanotubes. *Colloids Surf. A Physicochem. Eng. Asp.* **2016**, *495*, 118–124. [CrossRef]

56. Nam, I.W.; Choi, J.H.; Kim, C.G.; Lee, H.K. Fabrication and design of electromagnetic wave absorber composed of carbon nanotube-incorporated cement composites. *Compos. Struct.* **2018**, *206*, 439–447. [CrossRef]

57. Jiang, S.; Zhou, D.; Zhang, L.; Ouyang, J.; Yu, X.; Cui, X.; Han, B. Comparison of compressive strength and electrical resistivity of cementitious composites with different nano- and micro-fillers. *Arch. Civ. Mech. Eng.* **2018**, *18*, 60–68. [CrossRef]

58. Song, H.-W.; Lee, C.-H.; Ann, K.Y. Factors influencing chloride transport in concrete structures exposed to marine environments. *Cem. Concr. Compos.* **2008**, *30*, 113–121. [CrossRef]

MDPI

St. Alban-Anlage 66

4052 Basel

Switzerland

Tel. +41 61 683 77 34

Fax +41 61 302 89 18

www.mdpi.com

Applied Sciences Editorial Office

E-mail: applsci@mdpi.com

www.mdpi.com/journal/applsci

www.ingramcontent.com/pod-product-compliance
Lightning Source LLC
Chambersburg PA
CBHW051856210326
41597CB00033B/5920